Biosphere Reserves and Sustainable Development Goals 2

Book published within the framework of the Erasmus+ "EduBioMed"

This project has been funded with support from the European Union. This publication reflects only the author's view and the Agency and the Commission are not responsible for any use that may be made of the information it contains.

Project number: 598924-EPP-1-2018-1-ES-EPPKA2-CBHE-JP

Territory Development Set

coordinated by
Angela Barthes

Volume 4

Biosphere Reserves and Sustainable Development Goals 2

Issues, Tensions, Processes and Governance in the Mediterranean

Edited by
**Bruno Romagny
Catherine Cibien
Angela Barthes**

WILEY

First published 2023 in Great Britain and the United States by ISTE Ltd and John Wiley & Sons, Inc.

Apart from any fair dealing for the purposes of research or private study, or criticism or review, as permitted under the Copyright, Designs and Patents Act 1988, this publication may only be reproduced, stored or transmitted, in any form or by any means, with the prior permission in writing of the publishers, or in the case of reprographic reproduction in accordance with the terms and licenses issued by the CLA. Enquiries concerning reproduction outside these terms should be sent to the publishers at the undermentioned address:

ISTE Ltd
27-37 St George's Road
London SW19 4EU
UK

www.iste.co.uk

John Wiley & Sons, Inc.
111 River Street
Hoboken, NJ 07030
USA

www.wiley.com

© ISTE Ltd 2023

The rights of Bruno Romagny, Catherine Cibien and Angela Barthes to be identified as the authors of this work have been asserted by them in accordance with the Copyright, Designs and Patents Act 1988.

Any opinions, findings, and conclusions or recommendations expressed in this material are those of the author(s), contributor(s) or editor(s) and do not necessarily reflect the views of ISTE Group.

Library of Congress Control Number: 2023943622

British Library Cataloguing-in-Publication Data
A CIP record for this book is available from the British Library
ISBN 978-1-78630-842-9

Contents

Presentation of the Authors of the Two Volumes. xiii

Introduction xxiii
Angela BARTHES, Catherine CIBIEN and Bruno ROMAGNY

Part 1. Process, Governance and Climate Change Across the Mediterranean 1

Introduction to Part 1 3
Catherine CIBIEN

Chapter 1. Biosphere Reserves in National Legislation and Public Policy 5
Catherine CIBIEN, Lahoucine AMZIL, Joelle BARAKAT, Antonio BONTEMPI, Pierre DOUMET and Maria Carmen ROMERA-PUGA

    1.1. Introduction 5
    1.2. The place of the "biosphere reserve" in national legislation 6
    1.3. The place of MAB national committees in national governments 9
        1.3.1. In Spain 9
        1.3.2. In France 11
        1.3.3. In Morocco 12
        1.3.4. In Lebanon 12
    1.4. The place of the "biosphere reserve" model in public policy 13
        1.4.1. In Spain 14
        1.4.2. In France 18
        1.4.3. In Morocco 23
        1.4.4. In Lebanon 25
    1.5. Discussion 26
    1.6. References 28

Chapter 2. The Emergence and Evolution of Mediterranean Biosphere Reserves in France 31
Catherine CIBIEN

2.1. Profound changes across first-generation sites (1977) 31
2.2. The recognition of local development projects promoting natural and cultural heritage. 34
2.3. References 37

Chapter 3. Perspectives on Mediterranean Biosphere Reserves. 39
Ken REYNA, Martí BOADA and Mchich DERAK

3.1. Close-up on the strengthening of the Mont Ventoux Biosphere Reserve's governance 39
    3.1.1. Introduction 39
    3.1.2. An iconic Mediterranean mountain 39
    3.1.3. Conserving and developing the assets of an exceptional area. 40
    3.1.4. Governance evolving with the times 40
3.2. Close-up on the Montseny Biosphere Reserve 42
3.3. Close-up on the Menorca Biosphere Reserve 43
3.4. Close-up on environmental education and SDGs, an opportunity for Mediterranean Biosphere Reserves 44
3.5. Close-up on the Intercontinental Biosphere Reserve of the Mediterranean. . . 46
3.6. References 50

Chapter 4. From the Ecological Quality Status Evaluation to the Knowledge Transferability. A Cross-cutting Experience in Montseny Biosphere Reserve 51
Sònia SÀNCHEZ-MATEO, Antoni MAS-PONCE and Roser MANEJA

4.1. Introduction 51
4.2. Mediterranean river basins as valuable and complex socio-ecosystems 52
    4.2.1. The evaluation of ecological quality status 52
    4.2.2. Knowledge transfer and environmental education 53
4.3. Study area: Montseny Biosphere Reserve 54
    4.3.1. Observatori Rivus, a cross-cutting project in Mediterranean river basins 55
    4.3.2. Sampling units 56
4.4. Research areas 58
    4.4.1. Biological monitoring 58
    4.4.2. Hydromorphological monitoring 62
    4.4.3. Physicochemical monitoring. 63
4.5. Environmental education, communication and training program 63
    4.5.1. Formal education 65
    4.5.2. Nonformal education 65

4.5.3. Informal environmental education. 65
4.6. A 15-year period implementing PROECA in the Montseny Biosphere
Reserve 66
4.7. Conclusion 68
4.8. Acknowledgements 69
4.9. References 69

Chapter 5. Do We Need to Choose Between Biodiversity, Industry and Tourism? A Metabolic Approach to Manage the Mediterranean Biosphere Reserve of Menorca 73
Alejandro MARCOS-VALLS

5.1. Introduction 73
5.2. Societal metabolism 74
5.3. MuSIASEM: integrating information from multiple scales to improve
participation and stakeholder engagement 75
5.4. The case of Menorca: a Mediterranean Biosphere Reserve with an action
plan to implement the sustainable development goals. 79
5.5. Menorca 2025. An Action Plan for the Menorca Biosphere Reserve 81
5.6. Metabolic performance of economic sectors in Menorca. Application of
the MuSIASEM approach 83
5.7. Discussion: do we need to choose between biodiversity, industry and
tourism? 85
5.8. Conclusion 88
5.9. References 89

**Chapter 6. The Jabal Moussa Biosphere Reserve (Lebanon):
A Private Association Initiative** 95
Pierre DOUMET and Joelle BARAKAT

6.1. Introduction 95
6.2. Rich by nature 95
6.3. A privately run biosphere reserve 98
6.4. International recognition 99
6.5. Administration led by socio-economic expectations 100
6.6. Efforts at increasing understanding and awareness of an exceptional
biodiversity 102
6.7. References 103

**Chapter 7. Understandings of Administration and Challenges to
Governance in the Arganeraie Biosphere Reserve (Morocco)** 105
Abdelaziz AFKER

7.1. Introduction 105
7.2. A biosphere reserve built around an iconic tree: the argan tree 105

7.3. An integrated approach to conservation and ecodevelopment 107
7.4. Participation-oriented administration 110
7.5. Regarding the research/education/management dialogue 111
7.6. References 112

Chapter 8. Reconciling Conservation and Sustainable Development: The Example of the Arganeraie 113
Abdelaziz AFKER and Saïd BOUJROUF

8.1. Introduction 113
8.2. The ABR, between conservation and sustainable territorial development: reconciling the irreconcilable 114
8.3. The complex challenges characterizing the ABR, or relevance and adaptation in conciliatory resilience. 116
8.4. Changes and scalable trends in the ABR: from project territories to a territorial project 117
8.5. The ABR, complexities and improved governance 119
8.6. References 120

Chapter 9. Patrimonialization and Challenges to Sustainable Development within the Arganeraie Biosphere Reserve 123
Wahiba MOUBCHIR and Saïd BOUJROUF

9.1. Introduction 123
9.2. The ABR: a territory valued for the endemism of its heritage resources 124
  9.2.1. The ABR: the context of its creation and functions 124
  9.2.2. The Arganeraie: a resource deposit under anthropogenic pressure 128
9.3. The ABR patrimonialization process 130
9.4. Paths of governance for the integrated management of the ABR 134
  9.4.1. The path of the contract 136
  9.4.2. The path of deliberation: consultation and concertation. 136
  9.4.3. The path of incentivization. 137
  9.4.4. The path of institutional rearrangement. 137
9.5. Conclusion 138
9.6. References 139

Chapter 10. The Oasis du Sud Marocain Biosphere Reserve: Challenges and Issues for the Durability of Water Resources 141
Lahcen AZOUGARH and Ahmed MOUHYIDDINE

10.1. Introduction. 141
10.2. Specificities of the Oasis du Sud Marocain Biosphere Reserve and the question of water 142
10.3. Regional development and the deterioration of water resources 144

10.4. Challenges and complexities of water resource management within the
OSMBR 145
10.5. Conclusion 146
10.6. References 147

Part 2. Issues and Case Studies in the Southern Mediterranean 149

Introduction to Part 2 151
Catherine CIBIEN

Chapter 11. Pesticide Residue in the Waters of the IBRM 153
Hind EL BOUZAIDI, Fatimazahra HAFIANE, El Habib EL AZZOUZI and
Mohammed FAEKHAOUI

11.1. Introduction 153
11.2. Materials and methods 154
  11.2.1. Materials used 154
  11.2.2. Methods used and procedures of analysis 156
11.3. Results and discussions 156
  11.3.1. Pesticide use 156
  11.3.2. Water compartment contamination risks in the upstream reaches
  of the Intercontinental Biosphere of the Mediterranean (IBRM) 158
11.4. Evaluation of the risks of pesticides to human health 160
11.5. Evaluation of the risks of pesticides for the environment 160
11.6. Conclusion 161
11.7. References 162

**Chapter 12. Forest Fires: Their Impact on the Sustainable
Development of the IBRM** 165
Rachid SAMMOUDI, Abdelkader CHAHLAOUI, Nadia MACHOURI, Lahoucine AMZIL,
El Habib EL AZZOUZI, Reda NACER, Kawtar JABER and Maya KOUZAIHA

12.1. Introduction 165
12.2. The phenomenon of forest fires in the northern provinces 167
12.3. Links between sustainable development and forest fires 169
12.4. Conclusion 170
12.5. References 171

**Chapter 13. The Social and Solidarity Economy and Biodiversity in
the Intercontinental Biosphere of the Mediterranean** 173
Hicham ATTOUCH, Soukaina BOUZIANI and Sonia ADERGHAL

13.1. Some framing of the concept of the social and solidarity economy 173
13.2. Development of natural resources in the Intercontinental Biosphere
Reserve of the Mediterranean (IBRM) and the SSE framework 175

13.3. The role of the SSE in the conservation and development of natural
resources. 177
13.4. Conclusion 180
13.5. References 180

Chapter 14. The Media Coverage of the Biosphere Reserve: Ambivalence Between the Protection of Nature and the Promotion of Territories. The Case of RBIM 183
Lahoucine AMZIL, Yamina EL KIRAT EL ALLAME and Faiza EL MEJJAD

14.1. Introduction. 183
14.2. Biosphere reserves: general background 184
14.3. The media environment around the biosphere reserve 188
    14.3.1. Place of the biosphere reserve in the media channel 190
    14.3.2. Role of media and biosphere reserve actors. 193
14.4. Representation of RBIM in the Moroccan media 195
    14.4.1. Role of stakeholders in the visibility and access to RBIM. 195
    14.4.2. Measures and strategies for improving the biosphere reserves 198
14.5. Concluding remarks 201
14.6. References 202

Chapter 15. Mid-Atlas Cedar Forests and Climate Change. 205
Driss CHAHHOU

15.1. Introduction. 205
15.2. General overview of climatic changes 206
    15.2.1. General aspects of climate change in the Mediterranean region. 206
    15.2.2. Effects of drought on trees and forest stands 207
    15.2.3. The role of forest stands with regard to the water retention capacity
    of soils 208
    15.2.4. Potential strategies for facing climate change. 209
15.3. The vulnerability of forests to climate change 209
    15.3.1. The vulnerability of Morocco's climatic context and foreseeable
    changes 209
    15.3.2. Deterioration, deforestation and transformation of forest habitats 210
    15.3.3. Cedar diebacks: an indicator of climate change. 211
15.4. Potential impacts of climate change on cedar forests 212
    15.4.1. Elements of the Atlas cedar vulnerable to climate change. 212
    15.4.2. Impact on the growing season and distribution area of the cedar 213
15.5. Conclusion 214
15.6. References 215

Chapter 16. The Legacy and Future of Conservation in El Kala National Park (Algeria) 219
Tarik DAHOU

16.1. Introduction. 219
16.2. Declinism, forest exploitation and management in the EKNP 221
    16.2.1. The legacy of declinism in the EKNP 225
    16.2.2. Uses of the EKNP's natural resources. 228
    16.2.3. The structure of rural revenue in the EKNP. 230
    16.2.4. Conservation for improved exploitation. 231
16.3. The spread of fishing and marine conservation in the EKNP. 233
    16.3.1. Lake and lagoon fishing. 236
    16.3.2. Maritime fishing. 238
    16.3.3. Trawling 240
    16.3.4. Seine purse fishing 242
    16.3.5. Drift net and longline fishing. 243
16.4. Marine conservation and declinist rhetoric. 244
16.5. Conclusion 246
16.6. References 247

Chapter 17. Social Representations of Biospheres and Sustainable Local Development in Bou Hedma (Tunisia). 251
Abdelkarim BRAHMI

17.1. Introduction. 251
17.2. Bou Hedma National Park. 252
17.3. Methodological research framework 253
17.4. Social representations of Bou Hedma National Park among the
surrounding population. 254
    17.4.1. Negative representations: "the park as problem" 255
    17.4.2. Positive representations: "the park as symbol of identity". 256
17.5. Discussion and interpretation 256
    17.5.1. Negative representations: "the park as problem" 256
    17.5.2. Positive representations: "the park as symbol of identity". 260
17.6. The cultural dimension. 260
17.7. The political dimension 260
17.8. The environmental dimension. 261
17.9. Conclusion 262
17.10. References. 263

**Chapter 18. Architecture and the Biosphere Environment in
Pedagogy: Design Visions for Sustainable Dwelling Communities**   265
Carla ARAMOUNY

    18.1. Introduction.   265
    18.2. Architecture and the environment.   266
    18.3. Jabal Moussa Biosphere Reserve and the studio's premise.   267
        18.3.1. Studio methodology and research in three scales of operation   268
        18.3.2. Student architectural proposals and results   283
    18.4. Conclusion   288
    18.5. References   289

List of Authors   291

Index.   295

Summary of Volume 1   297

Presentation of the Authors of the Two Volumes

Hannah Abou Fakher holds a bachelor's degree in Agricultural Studies and a master's degree in Environmental Engineering and Natural Resources. She has worked on various projects as an intern at the Nature Conservation Center at the American University of Beirut (AUB). Hannah is currently working with the International Federation of the Red Cross and Red Crescent Societies in the area of climate and resilience.

Mohammed Aderghal is a geographer; a higher education professor at the Faculty of Letters and Human Sciences of Mohammed V University in Rabat; Director of the Laboratory of Tourism Engineering, Heritage and Sustainable Development of the Territories (LITOPAD); President of the Association of Moroccan Geographers (ANAGEM); and founding member of the international joint laboratory (LMI) Mediterranean Terroirs: Heritage, Mobility, Change and Social Innovation (MédiTer). He works on the agro-pastoral and rural changes at the origin of territorial recompositions and the change in the relationship between society and natural resources in the mountains and in the pre-Saharan and Saharan regions.

Sonia Aderghal holds a PhD in Geography from the Faculty of Letters and Human Sciences (Mohammed V University in Rabat). A researcher and associate at the LITOPAD and LMI MédiTer, her work focuses on the territorial dynamics of Mediterranean coasts and their hinterlands. She focuses on themes relating to tourism, urbanization, new ruralities and the governance of territories.

Abdelaziz Afker is a water and forest engineer (1988, Salé, Morocco) and holds a master's degree in Human Resources and Organization Management (Nancy 2009). The focal point for the Arganeraie biosphere reserve at the Regional

Directorate of Water and Forests in Agadir, he has held several positions of responsibility related to management, studies and partnership for the benefit of natural resources.

Abdullah Aït L'Houssain is a PhD student in Geography at the Laboratory of Studies on Resources, Mobility and Attractiveness (LERMA) at Cadi-Ayyad University in Marrakech. The title of his thesis is "Climate Change, Migration and Sustainable Development". His areas of research are climate migration, social movement and territorial feminism.

Lahoucine Amzil is based at Mohammed V University, Rabat. He holds a dual education in Social Sciences (Mohammed V University, Faculty of Arts and Letters, 1998–2002) and in Ecological Sciences (Albert Ludwig-Freiburg University, Department of Geobotany and Forestry, 2002–2005). Lahoucine Amzil's doctoral research focuses on recent transformations of the traditional socio-economic system of the western High Atlas. Amzil's current research interests include sustainable approaches, migration processes and rural development.

Carla Aramony is an architect and assistant professor at AUB.

Nina Asloum is an associate research professor (HDR) in Education and Training Sciences at the National School of Agricultural Education in Toulouse. She is responsible for the training of teachers in Agricultural Education for Spatial Planning. Her research focuses on curricula for producing differently and on the socio-historical evolution of training curricula for agricultural education within the Toulouse UMR Education, Training, Work, Knowledge (EFTS) and for the theme Changes in Education and Training: Engagement, Interactions, Emancipation.

Hicham Attouch is a professor at Mohammed V University, Rabat, in the Faculty of Legal, Economic and Social Sciences.

Lahcen Azougarh is a researcher at the Environment, Development and Spatial Management Laboratory at Ibn Tofail University, Kenitra, Morocco.

Didier Babin is president of the French committee of UNESCO's Man and the Biosphere (MAB) Programme and was president of the International Coordinating Council from 2016 to 2018. He is a member of the French National Biodiversity Committee. He is currently responsible for the Post 2020 Biodiversity Framework EU Support project at Expertise France. A doctor in Geography, researcher at CIRAD and associate at the University of Quebec in Montreal, Didier Babin was part of the CBD secretariat for five years as head of the Biodiversity for Development and Poverty Eradication program and was a member of the United Nations technical team for the preparation of the Sustainable Development Goals.

He was also involved in the emergence of the Intergovernmental Science-Policy Platform on Biodiversity and Ecosystem Services (IPBES) as executive secretary of the IMoSEB process.

Joelle Barakat is a conservation manager at the Jabal Moussa biosphere reserve, Lebanon.

Angela Barthes is a professor at the University of Aix-Marseille and specializes in the field of environmental education and the development of rural territories. She leads the Education, Territories, Development, Society, Health team from the ADEF laboratory.

Laurent Bedoussac is an associate research professor (HDR) in Agronomy at the National School of Agricultural Education in Toulouse. He leads the training of teachers in horticultural production and viticulture. He conducts research in the UMR AGIR at the National Research Institute for Agriculture, Food and the Environment on the analysis of the performance and functioning of mixtures of species for the design of agroecological cropping systems.

Sylviane Blanc-Maximin is a doctor with an HDR in Education Sciences specializing in the links between territory and education. She is a teacher trainer at the University of Aix-Marseille and a member of the ADEF laboratory.

Martí Boada is a professor at the Autonomous University of Barcelona (UAB), a member of the Institute of Environmental Sciences and Technologies (ICTA), a doctor of environmental sciences, the author of over 100 popular science books and a pioneer in ecology.

Antonio Bontempi is the project manager for EduBioMed and a doctoral student at the department of Geography at UAB. He holds a joint European MSc in Environmental Studies (2017) and was awarded another MSc in Building Engineering from the University of Bergamo (Italy 2013). He has been a visiting professor at several universities: the University of Columbia in New York (2016), the Technical University of Hamburg (2016) and the Instituto Superior Técnico de Lisboa (2012). Antonio's interests include protected areas, socio-environmental studies, ecological economics, political ecology, planning and territorial development.

Meriem Bouamrane is an environmental economist and Head of the Biodiversity and Ecology section within the Division of Ecological and Earth Sciences in the Man and the Biosphere (MAB) Programme. She is responsible for research and training programs on the access to and use of biodiversity, participatory approaches and consultation, using biosphere reserves as research and demonstration sites. She is the biodiversity focal point within UNESCO, as well as for IPBES and Future Earth.

Éliane Bou Dagher is a member of the Faculty of Sciences at Saint-Joseph University in Beirut (USJ).

Magda Bou Dagher Kharrat is a professor at the Faculty of Sciences at USJ in Beirut. Magda is the author of numerous scientific publications on Lebanon's biodiversity and the countless threats it faces. She is also the founder and president of the NGO Jouzour Loubnan.

Saïd Boujrouf is a professor of Geography, director of the Laboratory for Studies on Resources, Mobility and Attractiveness (LERMA) and co-director of the LMI MédiTer at Cadi-Ayyad University, Marrakesh. He studies land use planning and territorial development in Morocco.

Soukaina Bouziani, after finishing her baccalaureate studies in Physical Sciences, turned to Economics and Management Studies, successfully obtaining her license. Following short professional experiences in administration during and after her studies, Soukaina chose to complete her academic career with a master's degree in Governance, Territorial Planning, Local and Regional Development and Resource Management. In her final project, she addressed the themes of sustainable economy and the promotion of biodiversity in the Moroccan part of the Intercontinental Biosphere Reserve of the Mediterranean (IBRM).

Abdelkarim Brahmi is a lecturer at the Higher Institute of Applied Studies in Humanities at the University of Gafsa. He holds a doctorate in Heritage Sciences from FSHST, University of Tunis. He is a member of the Laboratory of Maghreb Studies at the same faculty.

Driss Chahhou is a professor at the Faculty of Sciences of Mohammed V University, Rabat.

Abdelkader Chahlaoui is a professor at the Faculty of Sciences of Moulay Ismail University in Meknes, head of the Valorization of Natural Resources Management research team and author and co-author of 100 scientific articles in peer-reviewed journals. He is also an expert for national and international bodies and initiator of scientific research projects.

Véronique Chalando is a doctoral student in the field of education for sustainable development; her area of research is agroecological knowledge. A certified professor of agricultural education in ecology and biology, she now works in agronomy at the University Institute of Technology of the University of Aix-Marseille.

Mikaël Chambru is a lecturer in Information and Communication Sciences at Grenoble Alpes University and researcher within the laboratory for excellence (Labex) Innovation and Territorial Transitions in the Mountains (ITTEM) and the Research Group on Communication Issues (Gresec). His work focuses on the methods for publicizing science in mountain territories, looking at mediation, communication and popularization of science, as well as the controversies and public challenges regarding Alpine societal issues in the context of a general demand for an ecological transition.

Catherine Cibien is the director of the French committee of the MAB Programme of UNESCO. She supervises the national network of biosphere reserves, including a set of working groups and projects relating to the implementation of the Sustainable Development Goals (biodiversity, education, communication, commitment of actors, etc.); acts as liaison with the network world; and takes part in the bodies of the Programme at UNESCO. She co-supervises the MAB masters at the University of Toulouse and chairs the Scientific Council of the Cévennes National Park biosphere reserve.

Cécilia Claeys is an associate research professor (HDR) of sociology at the University of Perpignan and a member of the Population–Environment–Development Laboratory. A sociologist, she conducts interdisciplinary research combining human sciences and life sciences. Her preferred fields of study are in mainland and overseas France. Her work focuses on the implementation of public environmental and risk prevention policies and their (non-)acceptance by populations and territorial actors.

Tarik Dahou is a research director at UMR 208 PALOC, IRD, National Museum of Natural History, Paris.

Mchich Derak is a forestry engineer at DREFLCD, Rif, Morocco.

Pierre Doumet is co-founder and president of the Association for the Protection of Jabal Moussa and manager of the Jabal Moussa biosphere reserve, Lebanon.

El Habib El Azzouzi is a professor at the Scientific Institute of Mohammed V University, Rabat, specialist in environmental chemistry and author and co-author of around 20 scientific articles in peer-reviewed journals. He is also a member of the Geo-Biodiversity and Natural Heritage Laboratory (GEOPAC) at ISR, a member of the management board of Mohammed V University and president of the association of UM5 laureates.

Hind El Bouzaidi is a member of GEOPAC, Mohammed V University, Rabat.

Wassim El-Hajj is an associate professor of Computer Science at AUB. His research focuses on the areas of wireless communication, network security and machine learning. His research activities have resulted in over 90 publications in reputable journals. He is a frequent speaker and technical reviewer at leading international conferences.

Yamina El Kirat El Allame obtained a PhD in Minority Identities, Languages and Cultures. She is an international advisor and consultant in the field of higher education. She is the coordinator of the AMAS project Erasmus Inter Africa Mobility in Morocco. El Kirat is also a coordinator of the Culture, Language, Education, Migration and Society Research Laboratory and of the Studies in Language and Society doctoral program at the Mohammed V University in Rabat in Morocco. For over 25 years, El Kirat has taught and conducted research on the disappearance of languages, cultural representations and attitudes, the analysis of political discourses and minority cultures, languages and identities. She is currently the coordinator of the UniMed Network on Migration.

Faiza El Mejjad obtained a bachelor's degree in English Studies at USMS Beni Mellal before embarking on a professional bachelor's degree in Tourism Engineering at Cadi-Ayyad University in Marrakech. El Mejjad's passion for the environment, space and territories led her to opt for the UM5R master's program Governance, Land Use Planning, Local and Regional Development and Resource Management and the Rural Tourism Heritage option to stay in touch with sustainability and biodiversity. Her master's thesis was a comparative study of media coverage of biosphere reserves in Morocco and Lebanon.

Mohammed Faekhaoui is a member of GEOPAC, Mohammed V University, Rabat.

Bruno Garnier is a university professor at the University of Corsica. The author of 60 publications, he continues his research with a focus on the right to education and cultural diversity in education. Within the laboratory Places, Identities, eSpaces and Activities (CNRS/University of Corsica), he is responsible for the team Identities, Cultures, Heritage Processes (ICPP).

Guillaume Gillet is a research engineer in agricultural and rural training at the National School of Agricultural Education in Toulouse and an associate researcher at UMR Innovation in Montpellier. He coordinates the training of teachers in agricultural equipment sciences and techniques. He conducts his research on support and training for actors who innovate in agriculture with a view to the agroecological transition.

Fatimazahra Hafiane is a member of the GEOPAC, Mohammed V University, Rabat.

Moustapha Itani holds an MSc in Ecosystem Management from AUB. After graduating, he became involved in many projects at the Nature Conservation Center at AUB. Moustapha is currently pursuing his doctoral research assessing the ecological impacts of pastoralism on high mountain landscapes as part of the Interdisciplinary Environmental Science program at the University of Helsinki.

Salma Itsmaïl is a doctoral student in Education Sciences at the University of Corsica and a member of the Places, Identities, Spaces and Activities laboratory (CNRS/University of Corsica). She works on the argan tree biosphere reserve.

Kawtar Jaber is a member of the Faculty of Sciences at USJ, Beirut.

Rhéa Kahalé is a member of the Faculty of Sciences at USJ, Beirut.

Maya Kouzaiha is a member of the Faculty of Sciences at USJ, Beirut.

Jean-Marc Lange is a professor of Education and Training Sciences at the University of Montpellier. His research, carried out from a curricular point of view, aims at the development of coherent and acceptable curricula for the formal education actors in the fields of transversal education. His work also focuses on the issues and orientation of educational policies and their historical roots. He is the author of numerous publications and co-authored the *Dictionnaire critique des enjeux et concepts des "éducations à"* (Critical Dictionary of the Issues and Concepts of "Education For") with Angela Barthes and Nicole Tutiaux-Guillon, published by L'Harmattan in 2017.

Nadia Machouri is a professor at the Mohammed V University in Rabat and specializes in Physical Geography. She is the author and co-author of around 40 publications in national and international journals and doctoral training coordinator in the field of Environmental Management and Sustainable Development. She is also a member of the Center for Studies and Research on Man, Spaces and Societies, having participated in eight national and international research projects.

Roser Maneja is the coordinator of the EduBioMed project. She is a doctor of Environmental Sciences (UAB 2011) and currently Deputy Director of Research and Senior Researcher at the Forest Science and Technology Center of Catalonia (CTFC). She is associate professor in the Geography department of UAB and a member of the Applied Geography Research Group and Visiting Professor at Centro de Investigaciones en Geografía Ambiental (CIGA), Mexico.

Alejandro Marcos Valls is a PhD candidate at ICTA at UAB. He holds an MSc from Uppsala University in Sustainable Development (2014) and two bachelor's degrees, one in Environmental Science (2010) and the other in Media and Communication (2011), both from the University of Barcelona. His current research focuses on integrated scenarios for island economies relying on a metabolic approach to cope with complexity in decision making and to promote participation and deliberation. Alejandro Marcos-Valls acknowledges having received financial support from the Spanish Ministry of Science, Innovation and Universities through the "Maria de Maeztu" Program for Units of Excellence (CEX2019-000940-M).

Anthoni Mas-Ponce carries out research activities mainly related to the monitoring of the ecological condition of Mediterranean river ecosystems and global change. He is currently preparing his PhD at the ICTA at UAB on the creation of new ecological indicators to assess the effects of global change in Mediterranean river basins.

Wahiba Moubhir is an accredited higher education professor in the department of Languages and Human Sciences at the Ecole Normale Supérieure in Marrakech. She is a permanent member of the Laboratory for Studies on Resources, Mobility and Attractiveness in the Faculty of Letters and Human Sciences of Marrakech and an associate member of the Center for Studies, Evaluation and Educational Research (CEERP) of the Cadi-Ayyad University of Marrakech.

Ahmed Mouhyiddine is a Doctor of Geography at Ibn Tofail University, Kenitra (Morocco), and in the Faculty of Letters and Human Sciences, Mohammed V University, Rabat.

Reda Nacer obtained a degree in Geography at the USMBA and a diploma in Business Management before embarking on a master's degree at the UM5R in Governance, Regional Planning, Local and Regional Development and Resource Management. His master's thesis focused on the intercontinental biosphere reserve of the Mediterranean and the subject of wood energy in the province of Chefchaouen.

Salma Nashabe Talhouk is a lecturer at the department of Landscape Design and Ecosystem Management at AUB. His work focuses on the promotion of community-based natural resource management; it explores digital technologies for nature conservation, highlights cultural ecosystem services and looks at the use of native and ecologically adapted plants in towns.

Nivine Nasrallah holds an MSc in Phyto-ecology from the Faculty of Sciences of the Lebanese University. After graduating, she worked as an intern, writing papers and articles at the Nature Conservation Center of AUB. Nivine is currently pursuing a second master's degree in plant protection at the Faculty of Agronomy and Veterinary Medicine of the Lebanese University.

Ken Reyna is the director of the Regional Natural Park and coordinator of the Mont Ventoux biosphere reserve, France.

Bruno Romagny is an economist, a research director (HDR) at the Institut de recherche pour le développement (IRD, Research Institute for Sustainable Development) and a member of the Population–Environment–Development Laboratory (LPED), UMR 151, at the University of Aix-Marseille. Together with Saïd Boujrouf, he co-directed the LMI MédiTer (2016–2021). The issues of access to and uses and modes of appropriation of renewable resources, as well as the difficulties raised by their concerted management and their valorization at a local scale, constitute the heart of his work.

Maria Carmen Romera-Puga is a socio-environmental science researcher interested in the rural commons, community protected and conserved areas, and political ecology issues in the Mediterranean. She is a PhD student at the ICTA at UAB (LASEG research group) and works on ICCAs, *agdals*, inclusive environmental governance and biosphere reserves in Morocco.

Rashid Sammoudi is a PhD research engineer at the Scientific Institute of Mohammed V University of Rabat, a specialist in ecology and environment and the management of natural resources and author and co-author of 10 Scopus scientific articles. A forest engineer and former head of water and forests, he has 20 years of experience in the management and valorization of Moroccan forest resources.

Sònia Sànchez-Mateo holds a doctorate in Environmental Sciences (UAB 2010) and a degree in Biology (UAB 2001). She is a researcher at the ICTA at UAB, a member of the Applied Geography Research Group at UAB and technical manager of Fundació Rivus, a public foundation promoting research, environmental education and dissemination for the conservation of river systems. She also coordinates the Observatori Rivus project, a long-term monitoring project that assesses socio-ecological indicators in Mediterranean river basins. Her expertise concerns riparian vegetation, water quality indicators, protected natural areas, global change and biodiversity.

Melki Slimani holds a doctorate in Education from the University of Montpellier and the Higher Institute of Education and Continuing Education in Tunis. His research focuses on the political dimension of environmental and development issues (*questions environnementales et de développement* (QEDs)) in the curricula. He teaches educational sciences at ISEAH, University of Kairouan, Sbeïtla, Tunisia.

Introduction

This book, *Biosphere Reserves and Sustainable Development Goals*, is organized in two volumes. The first volume specifically addresses scientific issues and educational practices in the Mediterranean region, whereas the second volume focuses on tensions, processes and governance. Case studies examine reserves on the southern, eastern and northern shores of the Mediterranean.

For the first time in the history of the panels of experts in charge of regularly assessing and synthesizing scientific knowledge on climate change and the erosion of biodiversity, approximately 50 Intergovernmental Panel on Climate Change (IPCC) and Intergovernmental Platform for Biodiversity and Ecosystem Services (IPBES) world specialists published a joint report[1] on June 10, 2021. In this document, they stated the imperative need to jointly address these two major phenomena, indicators of the different facets of the "great acceleration" of global change after the Anthropocene (Beau and Larrère 2018), in an attempt to provide responses commensurate with threats and challenges.

The interactions between climate and biodiversity have been more or less known for a long time, but these two areas are still too often treated separately, from the point of view of both research and public policies, or through the "greening" strategies in the activities of large private firms. In this report, researchers show that certain solutions presented as apparently good for the climate – such as single-species plantations of trees popular in carbon offsetting processes – can be harmful for biodiversity, the opposite being rarer. They remind us of the need to take into account these complex interactions at all the possible levels of action.

Introduction written by Angela BARTHES, Catherine CIBIEN and Bruno ROMAGNY.

1 See: https://www.actu-environnement.com/media/pdf/news-37685-rapport-atelier-giec-ipbes-climat.

In France, after being contacted by the municipality of Grande-Synthe in the north of the country (a municipality particularly exposed to climate change), the state council[2] set a new ultimatum to the public authorities. The government had a deadline of March 31, 2022, for implementing all the "useful" measures – without the specification of which – in order to bend the curve of greenhouse gas emissions and meet its goals. Although France has committed itself to reducing emissions by 40% by 2030 compared to 1990 levels and to reaching "carbon neutrality" by 2050, it is not taking concrete measures. Henceforth, administrative judges can control the action or inaction of the State in comparison with the recommended trajectories for the reduction of greenhouse gas emissions. This legal decision, the first of its kind in France, was taken after the High Council for the Climate, an independent body created at the end of 2018 and responsible for shedding light on the government's policies in this area, published a report[3] pointing out that the State was not making sufficient efforts to reach the 2030 goals. This was particularly relevant as the European Union had just raised its own climate goals: it had undertaken to reduce its net emissions by at least 55% by 2030 against the previous 40%, which will require an extra effort on the part of France.

Finance must become a key player in the ecological transition, accompanying its various dimensions (energy, agriculture, food, transport, etc.), all the more so given the fact that the banking sector is dependent on fossil fuel industries as never before. This is clearly shown by a study[4] published by a collective of associations (Friends of the Earth, the Rousseau Institute and Reclaim Finance). The study's abstract quotes the Banking On Climate Chaos 2021 report[5] according to which the 60 largest banks in the world granted EUR 3,393 billion financing to companies in the fossil fuel sector between 2016 and 2020. The in-depth study carried out by this collective of associations shows that the 11 main banks in the euro zone have accumulated a stock of over EUR 530 billion in assets related to fossil fuels, that is, 95% of their equity capital. Since January 2020, BNP Paribas, Société Générale, CréditAgricole and Natixis have granted over USD 17 billion in financing to the 30 most important companies in the world in terms of shale oil and gas exploitation. In this context, the devaluation of "fossil fuel assets"[6] held by the banks, which will

---

2 This is the body that advises the government on the preparation of bills and decrees. It is also the supreme administrative judge that settles the disputes relating to administrative acts.
3 See: https://www.hautconseilclimat.fr/wp-content/uploads/2021/06/HCC-rappport-annuel-2021.pdf.
4 See: https://gael-giraud.fr/actifs-fossiles-les-nouveaux-subprimes-notre-rapport/.
5 See: https://www.ran.org/bankingonclimatechaos2021/.
6 "Fossil fuel assets" are those contributing to the financing of exploration, exploitation and distribution (including transport, refining, etc.) for oil, gas and coal and to the production of electricity from these resources.

accompany the inevitable ecological transition, could produce significant turbulence or even generate a new global finance crisis. We would find ourselves in a situation comparable to the subprime crisis, when in 2008 the banking sector refused to open its eyes to the upcoming catastrophe and detonated a situation which could have been avoided, resulting in numerous bankruptcies and heavy socio-economic and human consequences.

While scientific warnings are old (Club of Rome report from 1972, etc.), those concerning the interactions between climate change and the rapid erosion of biodiversity have been taking shape for two decades, to the point of describing the current situation as the sixth major mass extinction crisis (Thomas et al. 2004). As the economist Laurent Eloi (2011) has pointed out:

> Our ecological crises reveal a paradox of knowledge and action: the considerable progress in environmental science over the past two decades has brought ever worse news about the state of ecosystems. [...] Even if the natural and physical sciences alert us – by pointing out significant areas of uncertainty – about the reality of ecological crises, they do not provide us with the means to transform attitudes and behaviors in human societies, societies which are responsible for planetary environmental change, behaviors and attitudes alone being capable of influencing its course. [...] In more provocative terms, one could say that, as far as environmental matters are concerned, the social sciences and humanities hold the key to solutions to the problems revealed by the hard sciences. This explains the necessary articulation between the two fields, if ecology does not want to be reduced to an ever more exact science of the contemplation of disasters.

Since the 1990s, various policies described as fostering "sustainable development" have supported mechanisms formalizing the links between the necessary management of biodiversity (Blandin 2009; Rockström et al. 2009) and local development, which must not only be inclusive and favor the reduction of social and territorial inequalities but also as neutral as possible in terms of greenhouse gas emissions. Among these systems, protected areas constitute one of the pillars of national and international nature conservation policies. To be effective, the size of protected areas must globally increase – for both terrestrial and marine ecosystems – and their interconnection has to be improved. However, it is also in terms of multi-stakeholder and multi-level governance that these reserves are expected to truly fulfill their role as an "open-air laboratory" during the ecological transition.

Recent developments systematically remind us of the importance of scientific approaches integrated with the social development needs from all globalized territories, as well as the need to articulate management modes in coordination with the economic and academic spheres. They also highlight the need to increase the engagement from citizens in all the available educational spheres (Cibien 2006).

In one of their last reports[7], experts from the United Nations Environment Programme (UNEP) and the Blue Plan emphasized that the rapidity of climate change in the Mediterranean basin is higher than the global trends. The Mediterranean has warmed by 1.5°C since the pre-industrial era, on average 20% faster than in the rest of the globe. This is what the recent synthesis of several hundred scientific studies has shown regarding the catastrophic consequences of global warming for the inhabitants of this world region, particularly when it comes to the supply of fresh water. By 2040, the Mediterranean region is expected to experience a temperature 2.2°C higher than at the end of the 19th century and locally 3.8°C higher by 2100 if serious mitigation measures are not taken expeditiously. "As a generalist diplomat, I am struck to learn that the Mediterranean is the second most impacted region in the world after the Arctic", summarized the Egyptian Nasser Kamel[8], Secretary General of the Union for the Mediterranean. "I understand that the state of the Arctic is important for the planet, but there are 500 million of us around this small lake…"

As elsewhere on the planet, the impact of climate change in the Mediterranean region has been exacerbated by other environmental problems relating to dynamics which have been underway for several decades, often supported by the public authorities. These phenomena have involved changes in land use, in particular those related to the rise of urbanization and mass tourism concentrated on the coast, and the intensification of agriculture, overfishing, declining biodiversity, soil degradation, desertification and pollution (atmospheric and aquatic environments). At the same time, these phenomena have led to increased competition between local actors and investors – sometimes foreign – for the access to resources related to water, soil (including beach sand)[9], forestry, fodder, etc.

When associated with demographic and socio-economic factors, climate change impacts water resources in various ways. Declining precipitation reduces surface and groundwater recharge. Extreme events (high floods, prolonged droughts)

---

7 UNEP/Mediterranean Action Plan and Blue Plan (2020). State of the Environment and Development in Mediterranean: Key messages [Online]. Available at: https://planbleu.org/soed/.
8 See: https://www.lemonde.fr/planete/article/2019/10/11/changement-climatique-les-scientifiques-s-inquietent-des-effets-sur-le-bassin-mediterraneen_6015045_3244.html.
9 See: https://wedocs.unep.org/bitstream/handle/20.500.11822/28163/SandSust.pdf.

accelerate soil erosion and threaten many human lives, as do recurrent heat wave[10] episodes. Water quality also suffers an impact, in particular due to the deterioration of wetlands and the salinization of groundwater in coastal aquifers.

The agricultural sector in the Mediterranean being the largest water consumer, the risks of conflicts between users have increased. In connection with demographic growth and the expansion of metropolitan areas, food demand is bound to increase accordingly (both at the national level and for export), which will automatically increase the demand for irrigation. Technical solutions in terms of water saving, such as dripping (Kuper et al. 2017), only postpone or delay the overexploitation of underground aquifers and create other difficulties (soil pollution by plastics abandoned in the open field, etc.). Significant adaptations are expected to modify agricultural[11] practices, defend the traditional Mediterranean diet, support localized agricultural production systems or even reduce waste and encourage a change in diets in line with the functioning of global food markets.

In the Mediterranean, terrestrial biodiversity is undergoing many changes. In the countries of the northern shore (especially coastal ones), urbanization has suppressed or fragmented many ecosystems. Due to the decline in agricultural and agro-pastoral systems, wooded areas are increasing at the expense of these two sectors. For their part, the semi-natural ecosystems of the countries along the southern and eastern shores are threatened with fragmentation or disappearance, notably due to urbanization, deforestation, the overexploitation of wood (for heating, construction) and overgrazing. According to the mentioned UNEP report, adaptation options at the level of terrestrial biodiversity include the preservation of "natural flow variability in Mediterranean rivers and the protection of riparian zones, reduction of water abstraction, modified silvicultural practices, and the promotion of climate-wise landscape connectivity".

---

10 For over a decade, each new year has broken the temperature records of the previous one in many countries on all continents. The summer of 2021 caused alarm with the appearance of a "heat dome" in Western Canada which resulted in the hottest day ever recorded in this country on June 30, 2021, with a 49.6°C temperature recorded at Lytton. This was one of the coldest territories on Earth showing the usual temperatures of the Middle East! As if to reinforce the dramatic dimension of the phenomenon and adding to its symbolism, the following day, the village of Lytton was almost completely destroyed by the flames of fires caused by this unprecedented heatwave episode. Other large fires also marked the summer of 2021 around the Mediterranean, specifically in Turkey, Greece, Algeria, France, etc.

11 Towards a productive agriculture capable of feeding the planet differently, in a less carbon-intensive manner, less chemical, less intensive and more respectful of humans, soils and biodiversity and more water-efficient.

Furthermore, crucial socio-economic factors are currently hampering sustainable development in Mediterranean countries. In the first place, there is persistent poverty, for example, in certain marginal rural regions in Northern Africa and in the urban and periurban suburbs which accommodate the most disadvantaged populations (unemployed, landless peasants, migrants, etc.). These pockets of poverty are related to inequalities and the imbalance between men and women. In addition, cultural dimensions also have to be taken into account to ensure the success of climate adaptation and environmental resilience policies in the heterogeneous multicultural framework of the Mediterranean basin. These policies, also aiming to support the most vulnerable local communities, must imperatively take into account social issues such as justice, equality, education, the fight against poverty, social inclusion and redistribution.

Finally, at the scientific level and with a view to meeting the main challenges in the Mediterranean, UNEP and Blue Plan experts insist on the need to quickly fill in the gaps between countries in terms of data and knowledge. In particular, they aim to promote the development of high-level climate services. This essentially points to early warning systems: "Increased research is required for short- and medium-term projections, as well as large-scale programs at the Mediterranean level to face impending challenges."

Since 1971, the UNESCO Man and the Biosphere Programme, known by the abbreviation MAB, has fostered different articulation principles between the political, scientific and academic spheres in various territories. Theoretically, the declared goal is to reconcile sustainable development at the regional level with the protection of the environment and, more specifically, the conservation of biodiversity, with respect for the cultural values of all. Biosphere reserves, which are formalized by UNESCO, are considered privileged territorial spaces for experimenting with operating methods specifically adapted to the stated goals and cross-cutting issues.

Today, the UN biosphere reserve system supports the 2030 Agenda and its 17 Sustainable Development Goals (SDGs). It encourages the development of interdisciplinary research and relies on its worldwide network to spread its experiences, its approaches and know-how. The different global and local political scales are articulated, with various consequences on the reconfiguration of local political arenas; on specific modes of development regarding the renewal of bonds between knowledge, powers and institutions and renovated relationships between the scientific and educational spheres; and on the governance of territories.

Responding to a multidisciplinary requirement and to a citizen–actor–researcher dialogue, this work addresses the question of the articulations and tensions between biosphere reserves and SDGs in the Mediterranean area from the perspective of complexity and multi-referentiality. It was conducted within the framework of the

Erasmus+ EduBioMed[12] (2018–2021) European program, led by the Autonomous University of Barcelona (UAB). This program also engages five other universities: the University of Aix-Marseille in France, Mohammad V in Rabat and Cadi-Ayyad in Marrakech in Morocco, and Saint-Joseph University and the American University in Beirut in Lebanon. It is also established a partnership with several biosphere reserves in France (Mont Ventoux), Spain (Montseny), Morocco (the Arganeraie and the Intercontinental Reserve of the Mediterranean) and Lebanon (Jabal Moussa, Chouf), as well as with the MAB France and UniMed networks.

The book is structured in four distinct and complementary parts. The first part focuses on the multidisciplinary scientific issues of biosphere reserves as part of the implementation of the SDGs in the Mediterranean. The second part focuses on educational and civic prescriptions and practices relating to biosphere reserves. The third part (Volume 2) guides reflection on the governance processes in climate change on both sides of the Mediterranean. Finally, the last part (Volume 2) engages the actors to express themselves on local issues specific to the south of the Mediterranean. This book has the particularity of resolutely taking part in the cross-dynamics of Mediterranean biosphere reserves and is enriched by the points of view from Lebanese, Tunisian, Moroccan, French and Catalan actors involved in the field and in the academic sphere.

This book intends to make its contribution to the specificities of the Mediterranean biosphere reserves, analyzing the articulations and tensions between the development of hinterland territories and environmental protection. It questions to what extent the challenges of the erosion of biodiversity and governance difficulties encountered in the Mediterranean biosphere reserves are actually taken into account. It also contributes to formalizing the links between academic research, citizens and biosphere reserve managers without concealing issues such as knowledge conflicts, citizen involvement and educational realities in a problematic framework which ends up being more complex than suspected at first sight.

References

Beau, R. and Larrère, C. (eds) (2018). *Penser l'Anthropocène*. Presses de Sciences Po, Paris.

Blandin, P. (2009). *De la protection de la nature au pilotage de la biodiversité*. Quae, Versailles.

Cibien, C. (2006). Actualités de la recherche. Les Réserves de biosphère : des lieux de collaboration entre chercheurs et gestionnaires en faveur de la biodiversité. *Natures Sciences Sociétés*, 14, 84–90.

---

12 See: https://www.edubiomed.eu/fr/.

Eloi, L. (2011). Économie du développement soutenable. Quelle place pour l'économie dans la science de la soutenabilité ? *OFCE*, 120, 7–12.

Kuper, M., Ameur, F., Hammani, A. (2017). Unravelling the enduring paradox of increased pressure on groundwater through efficient drip irrigation. In *Drip Irrigation for Agriculture: Untold Stories of Efficiency, Innovation and Development*, Venot, J.P., Kuper, M., Zwarteveen, M. (eds). Routledge, London.

Rockström, J., Steffen, W., Noone, K., Persson, Å., Chapin III, F.S., Lambin, E.F., Lenton, T.M., Scheffer, M., Folke, C., Schellnhuber, H.J. et al. (2009). A safe operating space for humanity. *Nature*, 461, 472–475.

Thomas, C.D., Cameron, A., Green, R.E., Bakkenes, M. (2004). Extinction risk from climate change. *Nature*, 427, 145–148.

PART 1

Process, Governance and Climate Change Across the Mediterranean

Introduction to Part 1

At the heart of the first part of this work lie questions of governance. The diversity and multiplicity of participants, the complexity of the play of actors, and their organization are decisive factors for the management of resources and territories.

Part 1 of this book will present the diverse situations visible in four countries included in the Network of Mediterranean Biosphere Reserves. How have they organized themselves in order to implement the model of flexible management represented by the UNESCO Biosphere Reserves? Have these reserves been factored into their laws and public policies? How are they funded? These reserves are the object of a panoply of institutional situations across the Mediterranean region, in Spain, France, Lebanon and Morocco.

How do the actors on the ground operate in practice? What problems and pressures do they face in their attempt to consolidate the conservation and development, by populations in sometimes precarious situations, of a rich, original and threatened biodiversity? How can they conserve the natural and cultural values of a region and make them known, while taking advantage of them? What kinds of balance can be found, conserved or rediscovered, and by what means? What positions can research, education and participation occupy?

A "visit" to several biosphere reserves across the four countries will allow the reader to understand some of these situations, many having the potential to become objects of transdisciplinary research in sustainability science.

Introduction by Catherine CIBIEN.

This voyage will lead the reader notably to the biosphere reserves of Mont Ventoux in France and of Montseny in Minorca, Spain, representing the north of the Mediterranean, showing how local actors organize themselves, promote and develop their natural riches, and showing the compromises and synergies they seek between different kinds of human activity.

In the south of the Mediterranean, the voyage will continue in the Jabal Moussa Biosphere Reserve in Lebanon, where the recognition, protection and development of heritage is used as a means towards local development. Then, in Morocco, the reader will approach the complexity of governance and its evolution with regard to the Arganeraie Biosphere Reserve, looking also at the multiple challenges tied to the exploitation of precious argan oil. This will also help measure the importance of public policy and matters of governance for the management of a crucial resource – water – in the Oasis du Sud Marocain Biosphere Reserve.

1

Biosphere Reserves in National Legislation and Public Policy

1.1. Introduction

Biosphere reserves, beginning in the 1970s, are land management apparatuses established within the framework of UNESCO's intergovernmental scientific program Man and the Biosphere (MAB).

Initially they were sites devoted to research and monitoring and regional bases for the program (Jardin 2021), but today they propose an integrated approach, securing a conservation of biodiversity and biological resources tied to the sustainable development of populations and based on knowledge in its broadest sense: scientific tracking and research, allowing for better demarcation of these areas' challenges, with regard to all of their natural and human components, and for an improved ability to respond to these challenges in an appropriate manner; and education, training, public awareness and local participation, allowing for a broader understanding of these challenges by a greater number of people. They stand as demonstrations of the implementation of the sustainable development objectives set by the United Nations for 2030 (UNESCO 2016).

For a color version of all the figures in this chapter, see www.iste.co.uk/romagny/biosphere2.zip.

Chapter written by Catherine CIBIEN, Lahoucine AMZIL, Joelle BARAKAT, Antonio BONTEMPI, Pierre DOUMET and Maria Carmen ROMERA-PUGA.

Since 1995, the functions and criteria that govern biosphere reserves have been formulated within texts officially endorsed by the UNESCO General Assembly: the Seville Strategy for Biosphere Reserves, and the Statutory Framework of the World Network of Biosphere Reserves (UNESCO 1996). A large number of countries have now established biosphere reserves, organized within the World Network, 727 in 131 countries as of 2021. As a formal designation from UNESCO, the "biosphere reserve" is a management model, but it is also considered an internationally prestigious label leading to patrimonialization. Which part is played by the former characteristic in the engagement of nations with the MAB framework?

This chapter aims to determine whether four Mediterranean nations (Lebanon, Morocco, Spain, France) have adopted this management model and whether, 25 years after their introduction, they have integrated the above texts into their legislations and institutions and, if so, how they have put them into practice. Which ties have they managed to establish between the MAB program, often associated initially with the scientific community, and public policy? How does this influence ways of managing and conserving biological resources and biodiversity? And how does it influence the management of lands and local development?

1.2. The place of the "biosphere reserve" in national legislation

Biosphere reserves are mentioned in the laws of the two European countries, but not in those of Morocco or Lebanon (Table 1.1).

Country	Biosphere reserves mentioned in national law	Date of introduction into the law	Mentions
Lebanon	No	–	–
Morocco	No	–	–
Spain	Yes	2007 and 2015	Law 42 on Natural Heritage and Biodiversity (articles 3.31, 65, 66 and 67) Updated in law 33
France	Yes	2016	Law for the reconquest of biodiversity (article 66)

Table 1.1. *Mentions of biosphere reserves in national laws*

Lebanese law makes no mention of biosphere reserves as such. Designated by UNESCO, they are only protected as far as local jurisdiction enacts a relevant juridical framework and implements the law appropriately. In Lebanon, laws protecting natural sites date from 1939 and, when implemented, offer a broad measure of protection for numerous natural and cultural sites. A new law on protected areas was enacted in 2019 under the supervision of the Minister of Environment. It encompasses public and private protected areas and details the planning and management required for initial designation as a protected area, and for the preservation of this status.

The particular value of biosphere reserves is that, rather than settling for the protection of the flora and fauna of a particular area, the conservation of a biosphere reserve must promote the ecosystems which profit the greatest possible number of rural and urban dwellers. To realize this new approach, an essential first step is the delimitation of all of the public and private land holdings and the implementation of land use regulations. Without these twin parameters of delimitation and land use regulation, most conservation activities are doomed to fail.

In Morocco, the law makes no explicit mention of biosphere reserves, and offers no specific juridical instruments for the adaptation and implementation of the MAB program in legislation and national policy. In consequence, the regulations incorporating Morocco's commitments to the MAB program are limited to the signing and ratification of international accords/conventions (such as the Convention on Biological Diversity in 1996) and to the official support given to each biosphere reserve for its designation, framework and action plan, and periodical reviews to be sent to UNESCO.

But on the occasion of a revision to Law 22-07[1] on Protected Areas enacted in August 2010 (BO August 19, 2010), the redesigned juridical framework adapted the criteria for international recognition to the specific conditions of Morocco. This development aims to involve administrations, regional authorities and organizations, and NGOs in the management and sustainable development of protected areas. Law 22-07 allows us to consider the management of protected areas as a tool for the conservation and development of biodiversity and of local sustainable development. But the lack of an implementing decree defining the responsibilities of actors and the raising of financial and human resources reduces the profile of Law 22-07.

---

1 Dahir no. 1-10-123 of 3 chaabane 1431 (July 16, 2010) promulgating Law no. 22-07 relating to protected areas. This law does not mention biosphere reserves among the protected areas recognized in Morocco.

The two European countries, Spain and France, offer systems reliant on the UNESCO texts:

– In Spain, the MAB has been strongly supported by its institutions, thus reinforcing the Spanish Network of Biosphere Reserves and inscribing it in law from 2007. Law 42/2007 of December 13 on Natural Heritage and Biodiversity takes into account the existence of biosphere reserves (articles 3.31, 65, 66 and 67). Article 65 indicates that the Spanish Network of Biosphere Reserves constitutes a defined and recognizable subgroup of the World Network of Biosphere Reserves. The text takes into account the objectives of UNESCO's MAB program and provides for the possibility for biosphere reserves to include other protected areas. Their characteristics are notably based on the criteria of article 4 of the Statutory Framework of the World Network (UNESCO 1996): zooming, governance, etc. The 2007 law was updated by Law 33/2015 of September 21, 2015.

– In France, the law mentions biosphere reserves more belatedly. The 2016 Law for the Reconquest of Biodiversity (article 66) considers biosphere reserves as contributing to the objective of sustainable development. It stipulates that:

> in application of Resolution 28C/2.4 of the UNESCO General Assembly which approved the Seville Strategy and adopted on 14 November 1995 a Statutory Framework of the World Network of Biosphere Reserves, regional authorities, their groupings, the 'syndicats mixtes'[2] mentioned in Chapter VII of the fifth part of the general code of regional authorities, the national public associations and establishments administrative of parks may put into place a biosphere reserve. A biosphere reserve contributes to the objective of sustainable development defined in II of article L. 110-1 of the present code. The national strategy for biodiversity favors the development of biosphere reserves in mainland France and its overseas territories and dominions.

This text therefore does not define the biosphere reserve as a new category of protected area, with France already having a full range of such areas. It allows them to be organized in reliance on one or more administrative structures (in support of protected areas of different types and regional authorities) and associations combined according to the zoning model specific to biosphere reserves. In contrast to Spain, the law makes no reference to the network(s) constituted by the reserves, at the national or international levels.

---

2 In France, there are joint ventures between different public authorities.

1.3. The place of MAB national committees in national governments

Country	Creation date	Form as of 2021
Lebanon	–	Committee presided by a scientific figure (CNRS) who maintains the link to biosphere reserve support structures.
Morocco	1995	Honorary committee, created in 1995, composed of volunteers and presided by a scientific figure, who represents the biosphere reserve during national and regional meetings. This committee operates in conjunction with the High Commission for Water and Forests, this commission being in charge of protected areas, NGOs and ANDZOA.
Spain	1975	Collegiate, interministerial, advisory body, with representatives from multiple departments of the General State Administration, from all of the autonomous communities whose land contains biosphere reserves, from the scientific sector, and from other categories of actor.
France	1973 (and 2015)	MAB committee integrated into the National Commission for UNESCO from 1973. Modified in 2015 with the creation of a specific legal body, the MAB France Association. This gathers biosphere reserves, national scientific institutions (CNRS, INRA, CIRAD, MNHN, IFREMER, IRD), the relevant ministers (Foreign Affairs, Environment, Research and Education), public and private partner organizations dealing with questions of biodiversity and sustainable development, researchers and individual experts. The president (also the representative of France at the MAB-ICC) is named by the Minister for Foreign Affairs.

Table 1.2. *Organization of national MAB committees*

1.3.1. *In Spain*

According to the historical reconstruction of Spain's National Parks Autonomous Agency (Organismo autónomo de parques nacionales (OAPN 2012)), the Royal Decrees 342/2007 of March 9 2007 and 387/2013 of May 31 currently govern the development of the MAB program's functions, as well as those of the Spanish MAB Committee, within the OAPN.

The Spanish National Committee for the MAB program, since its creation in 1975 under the aegis of the Spanish National Committee for Cooperation with UNESCO, has seen considerable development, of which several stages can be identified.

Between 1975 and 1987, the Spanish MAB Committee functioned as a work group within the framework of the Spanish National Committee for Cooperation with UNESCO. At this stage, its composition was essentially academic, despite representatives from the domains of management and administration being gradually integrated.

In 1987, with the creation of an MAB Committee Support Office within the General Directorate for the Environment of the Ministry of Public Works and Urban Development, the activities of the MAB Committee received a strong boost which, with time, became focused on the concept of the biosphere reserve (OAPN 2012). Consequently, a technical work group for biosphere reserves was created in 1992, marking the beginning of the formation of the Spanish Network. The development of this group's tasks culminated in the Spanish contribution to the 2nd World Congress of Biosphere Reserves in Seville in 1995, organized by UNESCO. The period from 1987 to 1996 can be seen as a period of mutual understanding and interest between the Spanish MAB Committee and the public institutions responsible for the environment, at both regional and state levels.

The Royal Decree 1984/1996 of August 2 1996, which made reference to the structure of the Ministry of Environment, assigned to the OAPN the function of coordinating and developing the MAB program (OAPN 2012). These coordination functions have been maintained following the Royal Decree 1130/2008 of July 4, relative to the organic structure of the Ministry of Environment and Rural and Marine Affairs. The Royal Decree 401/2012 of February 17, which developed the organic base structure of the Ministry of Agriculture, Food and Environment, did not modify the functions of the National Parks Autonomous Agency with regard to the development of the MAB program in Spain. Consequently, since 1996, the OAPN has supported a technical office to aid the development of the MAB Committee's activities. From 1996 to 2007, the MAB Committee continued to carry out its activities, notably through an increase in the number of Spanish biosphere reserves and the impetus given by Spain to the IberoMAB Network (a network gathering the nations of Latin America, the Caribbean and the Iberian Peninsula).

The Royal Decree 342/2007 of March 9 2007, governed the development of the MAB program's functions, as well as those of the Spanish MAB Committee and its advisory bodies (the Scientific Council and the Board of Biosphere Reserve Managers) under the aegis of the OAPN. This royal decree, which specified the manner in which the OAPN ought to carry out its functions, assured institutional support for the functioning of the MAB program in Spain and gave a new impetus to the development of the program's activities. Almost simultaneously, the inclusion of biosphere reserves in Law 42/2007 of December 13, on Natural Heritage and Biodiversity, provided regulatory backing and consolidated the Spanish Network of Biosphere Reserves.

The decree also specified the composition and functions of the Spanish MAB Committee, as well as its ties with the Agency, making the committee an effective instrument of institutional coordination for the biosphere reserves and for the transfer of initiatives to institutions. The Spanish Committee for the MAB program has since been a collegiate interministerial advisory body, with representatives from multiple departments of the General State Administration, from all of the autonomous communities whose land contains biosphere reserves, from the scientific sector, and from other categories of actor. The MAB Committee is equipped with two advisory bodies: the Scientific Council and the Board of Biosphere Reserve Managers, who are very active in the creation of initiatives and the putting in place of biosphere reserves in Spain.

Spain's investment in the MAB program is unique in that, on top of the Seville Congress, which provided biosphere reserves with their fundamental texts, the third World Congress of Biosphere Reserves was held in Madrid in 2008, and the international congress Seville +5 was held in Pamplona in 2000, the latter providing the recommendations which determine the modalities for the creation and functioning of transfrontier biosphere reserves still in place today.

1.3.2. In France

The MAB Committee was created in 1973. From 1973 to 2015, the MAB Committee functioned under the aegis and with the support of the French National Commission for UNESCO. At this stage, its composition was essentially academic, and with representation from the relevant ministries (environment, foreign affairs, education). From 1991, representatives from biosphere reserves have been associated with the MAB Committee. Its funding primarily came from the ministry responsible for the environment, the Ministries for Foreign Affairs and for Research and Education providing small supplementary budgets. In the first few years, leadership was provided by a researcher from the ORSTOM (today the IRD). From 1991, the MAB Committee was able to recruit a scientific secretary thanks to funding from the ministry responsible for the environment. From 2002, a CNRS researcher was also associated with the leadership of the MAB Committee.

With the development of the National Network of Biosphere Reserves, and notably with the arrival into this network of three new sites between 2012 and 2014, it became necessary to introduce a legal figure into the MAB Committee's leadership. This was done with the intention of giving the committee the profile it lacked, to allow for more ambitious projects and to better promote the "biosphere reserve" model within the French conservation and sustainable development landscape. This was done in 2015 with the creation of the association MAB-France. Its governance relied on the participation of its active members, large French research bodies and biosphere reserve support structures, associated members, ministers, institutional partners (notably the French Office for Biodiversity) and associations. Its administrative council is composed of 15 members, half from biosphere reserves and half scientists, elected for three years. Its staff is composed of two people (one director and one project manager) supported by a CNRS researcher.

Its principal budget is supplied by the French Office for Biodiversity as created in 2020, filled out by modest contributions from the ministries responsible for higher education and research and for foreign affairs, and on the basis of specific projects. The association stands as the French MAB Committee, and as such, manages candidacies and periodical reviews. It leads the French Network of Biosphere Reserves, work groups (for research, youth engagement, EDD (sustainable development education), forestry management, and communication) and projects relating to the development of the companies and actors of the lands in question (awards, eco-actors). It also jointly runs the MAB master's degree program at the University of Toulouse. The president of the MAB-France Committee, generally a recognized scientific authority, is named by the Ministry of Foreign Affairs. It represents France at the MAB-ICC, in conjunction with the UNESCO delegation from France.

1.3.3. In Morocco

The Moroccan MAB Committee was created in 1995, a little before the establishment of its first biosphere reserve (the Arganeraie Biosphere Reserve, declared in 1998). The Moroccan MAB Committee, honorary, is comprised of volunteers, normally researchers without administrative responsibility. The current president is in frequent contact with the National Agency of Water and Forests in Rabat and has a relationship with the ANDZOA (National agency for the Development of Oasis and Argan Zones) and the forestry administration. It is present in the majority of national debates and forums regarding biosphere reserves and in UNESCO international meetings, particularly at the MAB-ICC. As in other southern Mediterranean countries, Morocco has no specific juridical instruments for the adaptation and implementation of the MAB program into law or national policy. As in other southern Mediterranean countries, Morocco has no specific juridical instruments for the adaptation and implementation of the MAB program into law or national policy.

1.3.4. In Lebanon

The Lebanese Committee for the Man and Biosphere Program (LebMAB) is supported by the Lebanese National Centre for Scientific Research (CNRS). Initially composed of three members belonging to this body (the president, Professor Tohmé, and the secretary, Ghassan Jaradi), it was enlarged to encompass a representative from each biosphere reserve (Shouf, Rihane and Jabel Moussa) following the World Congress of Biosphere Reserves in Lima in 2016.

1.4. The place of the "biosphere reserve" model in public policy

The Management Manual for UNESCO Biosphere Reserves in Africa (2018), published by the German Commission for UNESCO, presents two categories of governance structure, one called the "authority model" and the other the "NGO model".

In the first model, the approach is top-down, the authority in question – a management unit dependent on a ministry or another authority – being primarily responsible for the conservation of the nation, and normally responsible for only the core area (and sometimes the buffer zone). When this is the case, it can be difficult for the authority to be active with regard to sustainable development, or within the transition zone.

In the second model, the "NGO model", the management committee, composed of multiple private and public institutions, serves as a platform for the meeting of interests and communities. This model is well adapted to acting in an advisory fashion, but generally has no direct power of intervention, and is often obliged to negotiate with other institutions to implement decisions made on its platform. It is also more oriented towards projects than towards management. Integrating management into this ensemble is more difficult.

Jardin (2017) proposes a different distinction between the existing structures and the ad hoc structures established for biosphere reserves at the time of their creation. The first category includes the body in charge of a protected area, extended or not, or of a part of an area, for example: a national part, a nature reserve or a national marine park. This body would have a direct power of intervention, but arrangements would need to be made to respond to the objectives of the biosphere reserve, notably with regard to the transition zone where the body would have no authority; a management committee would need to be added for consultation.

The case of an IUCN Category 5 protected area would need to be dealt with separately, as it would include the transition zone and thus correspond to the limits of the biosphere reserve.

In other situations, for example, when a perimeter is established on a geographical, ecological or cultural logic, it would include a public governing body adapted to its needs, such as a municipality with an added management committee and associations, or the government of an island (Menorca, the Isle of Man) in which special arrangements would be taken to respond to the objectives of the biosphere reserve. It would include a group of institutions and municipalities (a "syndicat mixte" in France; a public structure) or a group of partners, including associations (a private structure). It could also be comprised of public/private partnerships. In the case of a private structure, its role would be solely advisory.

1.4.1. *In Spain*

In 2021, the Spanish Network of Biosphere Reserves counted 52 sites spread over 16 of the country's 17 autonomous communities (OAPN 2021).

The practical implementation of the concept of the biosphere reserve has evolved in Spain, as it has done at the international level. The OAPN (2012) indicates that since the first biosphere reserves were named, from 1977 to 1992, all named biosphere reserves have previously been natural parks, national parks or nature reserves, with the exception of Urdaibai, which was declared in 1983 and was covered by a specific law. In 1993, Spain proposed the naming of two islands in their entirety: Menorca and Lanzarote, covered also by a specific law. They contained protected areas, and were presented as integrated management projects for a complex region, covering all sectors of production, all uses of the land and all of its inhabitants, and proposing an operational program based on participation, as well as objectives conforming to sustainable development. The impact of these two experiences on the approaches adopted at the Seville Congress in March 1995 (UNESCO 1996) was significant. From 1997 to 2006, 23 new areas were declared as biosphere reserves. Following this, the rate of declarations slowed while their integration into a network was reinforced, on the basis of the restructuring of the Spanish MAB Committee and of the support provided for the running of the reserves by the National Parks Autonomous Agency. The new situation offers solid support to the Spanish Network of Biosphere Reserves that is entering a period of intense activity, seeking to improve their reach and to adapt their situation to the requirements put in place at the Seville Congress in the cases of biosphere reserves lacking an adequate organizational structure.

The increase in requirements, however, has not diminished the number of areas aspiring to join the network. On the contrary, it seems to have the opposite effect, and there are many areas expressing their wish to be declared as biosphere reserves.

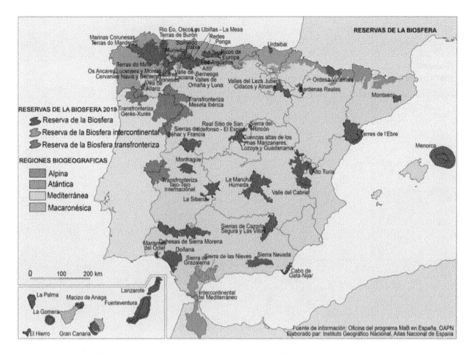

Figure 1.1. *Map of Spanish biosphere reserves (source: OAPN 2019)*

Santamarina Arinas (2015) puts forward an image of the legal situation of biosphere reserves in the official regulations of each autonomous community, and highlights the most important gaps. Figure 1.2 shows the mosaic of the different institutional systems of governance.

Around half of Spain's biosphere reserves (23) are governed directly by the environmental departments of the autonomous communities. The majority coincide with protected areas, following the schema dating from the end of the 20th century. This is the model chosen, for example, by the administrations of Andalusia, Extremadura, Madrid and Asturias. It has the advantage of strong budgetary security and is clear with regard to administration. However, in certain cases, examples of this model suffer from gaps with regard to the objectives of the MAB program in their concentration of the majority of their resources on the conservation of the protected area and logistical support for the core area, at the expense of socio-economic development. These spaces are governed by specific regulations that define their functions with regard to conservation and have often limited power with regard to the environmental domain, integrating only with difficulty the other sectors that play a pertinent role in a biosphere reserve: industry, tourism, culture, economy, etc.

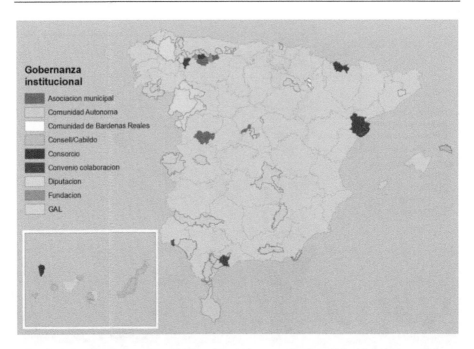

Figure 1.2. *Typology of institutional governance in Spanish biosphere reserves (Oñorbe 2018). In the legend, from high to low: municipal association; autonomous community; Bardenas Reales community; Consell/Cabildo; consortium; collaboration agreement; private council; foundation; Local Action Group (GAL)*

The case of the Sierre de las Nieves Biosphere Reserve in Andalusia ought to be highlighted: the biosphere reserve integrates a natural park into its territory. The management of the reserve and of the park is carried out in a coordinated but autonomous manner. The management body is articulated through a collaboration agreement that includes the relevant municipalities, an association and the autonomous community as parties responsible for the management of the natural park. The biosphere reserve's manager is an employee of the group of municipalities, responsible for the functions of the biosphere reserve on the ground, and who depends, in large part, on external opportunities for resources. It is a successful example that could well serve as a reference for other reserves.

In the same vein, in the cases of the Las Marismas del Odiel Biosphere Reserve (Andalusia) and the Cuenca Alta del Rio Manzanares Biosphere Reserve (Madrid), the traditional schema, superimposing a protected area and a biosphere reserve, is in place. However, at the request of UNESCO, it was necessary to extend their limits beyond the natural parks.

The model of governance evolved and integrated relevant municipalities and actors from outside the park.

Regional authorities (Diputaciones, Cabildos and Consells) are a governing mechanism involved in a quarter of the biosphere reserves in the Spanish network (11). They ensure higher levels of budgetary security, assignment of staff, and despite their legal powers being limited on the ground, they have a capacity for management. They serve as a liaising body between autonomous communities and municipalities, generally displaying good sectorial coordination.

The case of Montseny is another to be foregrounded, one which had to revise its management model following a process of extension beyond the limits of the natural park. The adopted solution consisted of the integration of the park into the framework of the provincial council within the reserve, rendering the director of the park naturally dependent on the manager of the reserve. Since then, the multifunctional model has displayed an exemplary performance. The park and its director are focused on running conservation efforts within the core area. The biosphere reserve, in coordination with those efforts, works on the other functions of the buffer and transition areas: namely socio-economic development and logistical support.

The other biosphere reserves are managed by an amalgam of local bodies which are not necessarily led by an administration (foundations, associations of municipalities, consortiums). One of their principal weaknesses is budgetary insecurity, but at the same time, they are extremely flexible and have a strong ability to adapt. This is the type of management chosen, for example, by the Castilla y Leon and the Aragón Biosphere Reserves (Ordesa Viñamala). It is worth mentioning in particular the Mariñas Coruñesas e Terras do Mandeo Biosphere Reserve (A Coruña) and the Allariz Biosphere Reserve (Ourense), managed by rural development associations under the supervision of the corresponding administrations.

A priori, the advantage of these more localized models is that they allow for greater contact between the managing bodies and the locals, favoring their participation and their involvement in land management. The main inconvenience of this model is the lack of budgetary and/or specific human resources with which to run the reserve.

Five autonomous communities have not yet developed a regulatory framework for biosphere reserves: Cantabria, Catalonia, Galicia, the Balearics and Navarre. This does not necessarily suggest that their biosphere reserves lack managing bodies and instruments.

A range of situations show that having regional regulations around biosphere reserves does not necessarily signify that there is *sufficient* regulation for the planning and organizing of reserve management. The content of the regulations still in place in Andalusia or Castilla-La Mancha, and, more recently in Madrid and even, to a degree, in Castile and León, perfectly illustrates what could be seen as another means of not materially fulfilling commitments.

The six remaining autonomous communities partially respect their commitments. There are important differences in content based on local circumstances present in each case and provide an interesting diversity, which may enrich it as a whole as well as the exchange of experiences. It is also true that, in most cases, it would be ideal to proceed towards adjustments or updates of differing scopes. But in general, we can say that the degree of conformity is higher with regard to the organization of biosphere reserves than in their management plans.

In surmounting these difficulties, the autonomous communities could fail to fulfill the task of coordination present at the level of the state, a coordination which is not a question of erasing reasonable differences, but of putting forward a modicum of shared guidelines to clarify doubts. In the meantime, this is not the only contribution that the legislator at the state level could make to help surmount the aforementioned problems, which are particularly complex where biosphere reserves coincide with certain national parks.

1.4.2. *In France*

France designated its first biosphere reserves from 1977, which remained undefined within French law until 2016, with the adoption of the Law for the Reconquest of Biodiversity. They thus represented a materialization on the ground of a scientific program, the MAB, and an internationally recognized designation.

MAB-France is nevertheless associated with different bodies relating to biodiversity and protected areas: it is part of the National Committee for Biodiversity of the Ministry of Environment, and of the National Conference for Protected Areas, under the aegis of the French Office for Biodiversity (OFB). This office was created in 2018 with a broad approach to conservation and biodiversity, as it relates to society. This implies a stronger support for the approaches of the MAB program than was previously seen, but a support which remains modest with regard to the challenges and possibilities of the network's development.

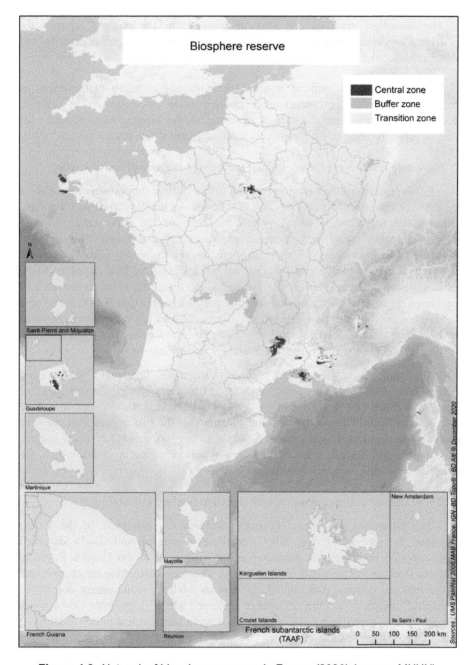

Figure 1.3. *Network of biosphere reserves in France (2020) (source: MNHN)*

The biosphere reserves are also integrated into the national system for biodiversity information (the National Inventory of Natural Heritage) run by the Muséum national d'histoire naturelle (MNHN).

In practice, each biosphere reserve relies on one or more management or leadership apparatuses: protected areas, municipalities, associations. The reserves receive no financing from the state or from regions for running costs as such, this funding rather filtering down through their management structures.

Biosphere reserves are thus delimited according to several different logics:

– Firstly, their perimeter may correspond to that of their support structure. This is the case for the national parks of the Guadeloupe Archipelago and the Cévennes, governed by a 2006 law which places the heart of the park within an area of free partnership for local communities, following a principle of ecological solidarity. The governance of the biosphere reserve coincides with that of the park (an administrating council comprised of institutions, local municipalities, socio-professional representatives, associations, etc.), and the management plan of the biosphere reserve is assimilated into the park's charter. It is important to note that the functions of the "new generation" of national park are very close to those of the biosphere reserve, their zoning being made up of two zones (rather than three): the heart of the park, which is regulated, and the partnership zone.

– The same principle of the superimposition of perimeters, bodies and management documents (and functions) applies to the *regional* national parks of Vosges du Nord (the French part of the German–French transfrontier biosphere reserve) and Mont Ventoux. Again, the functions of the two apparatuses are very similar. The regional national parks have no zoning systems. They often include protected areas (natural reserves or other types of protections) on which the biosphere reserve's zoning may be based.

These situations ensure stability with regard to funding and staffing:

– The Falasorma Dui Sevi Biosphere Reserve (an extension of the Fango Biosphere Reserve dating from 2019) constitutes a coherent and clearly identified subgroup with regard to the perimeter and management of the Corsica Regional National Park, in this case its maritime aspect. The decision-making structure is that of the regional national park, but it also has an advisory management committee belonging uniquely to the biosphere reserve which represents its various stakeholders, twinned with a scientific council and a management policy identifying specific objectives.

– The perimeter of the biosphere reserve may be laid out according to a geographical, ecological or socio-cultural logic which requires the collaboration of multiple structures which co-manage a coherent whole: this has been the case in the

Camargue Biosphere Reserve (the Rhône delta), one of the first French biosphere reserves (1977), since its extension in 2006. It is also the case for the Marais Audomarois Biosphere Reserve (the city of Saint-Omer and its wetlands), and the Îles et Mer d'Iroise Biosphere Reserve, where a regional natural park and a marine park co-manage the biosphere reserve. In each of these cases, each structure has decision-making power over its territory and applies its own management policies. Cooperation is defined for projects at the level of groups responding to specific funding. In the Fontainebleau et du Gâtinais Biosphere Reserve, a specific association ensures joint leadership between the relevant authorities, a regional natural park and the associations involved. There is also a dedicated scientific council.

– The large Bassin de la Dordogne Biosphere Reserve responds to the logic of a drainage basin. It is run by a public structure empowered to manage the rivers and aquatic spaces. This situation calls for the implementation of partnerships with other institutions or local authorities to deal with the range of missions required by biosphere reserves with regard to management (of agriculture, forestry, economic development, urbanism, etc.), education, etc. In the absence of specifically dedicated funding and of support from public powers, the biosphere reserve struggles to sufficiently enlarge its range of partnerships and thus to ensure leadership and presence at the scale required of such a large territory.

– The biosphere reserve may exceed the perimeter of its support structure (as is the case in the Biosphere Reserves of Luberon Lure, Mont Viso and the Gorges du Gardon). This necessitates political agreements with neighboring territories that are not always formalized, and sometimes prefigure the extension of the support structures' reach.

The absence of a proper budget, combined with these situations of superimposition, problematizes the public profile of biosphere reserves in France. On the contrary, the variety of institutional arrangements offers a great flexibility with regard to implementation, which is genuinely valuable, allowing for ease of adaptation to different challenges and for temporal changes, and offering spaces for dialogue (at the level of coherent bodies) beyond administrative and political divisions. This would certainly explain the new candidacies, despite the contributions of the "biosphere reserve" management model being little known and little supported by authorities (no public funding is currently allocated to any territory designated as a biosphere reserve). The particular values of the model, as well as its flexibility, are above all its ability to function within a network and in conjunction with the scientific and educational worlds, its international outlook and the renown of UNESCO.

	Dates of designation	Coordinating structure	Area (ha)	Populations
Commune de Fakarava	1977 Extended in 2006	Commune de Fakarava Biosphere Reserve Association	288,880	1,500
Camargue (Rhône delta)	1977 Extended in 2006	Camargue Regional Nature Park, in collaboration with the Syndicat Mixte Camargue Gardoise	160,000	From 110,000 to 220,000 (in summer)
Falasorma Dui Sevi Fango Valley	1977 Extended in 2019	Corsica Regional Nature Park	23,500	450
Cévennes	1985	Cévennes National Park	325,000	50,000
Îles et Mer d'Iroise	1988 Extended in 2012	Armorique Regional Nature Park and Iroise Marine Nature Park	200,000	1,400
Vosges du Nord Pfälzerwald	1989 Cross-border since 1998	Vosges du Nord Regional Nature Park in France, Natürpark Pfälzerwald In Germany	310,300	337,000
Mont Ventoux	1990	Mont Ventoux Regional Nature Park	80,368	30,000
Archipel de Guadeloupe	1999 Extended in 2013	Guadeloupe National Park	69,707	124,000
Luberon – Lure	1997 Extended in 2009	Luberon Regional Nature Park	244,645	170,000
Fontainebleau – Gâtinais	1998 Extended in 2009	Fontainebleau et du Gâtinais Biosphere Reserve Association	150,544	267,665
Bassin de la Dordogne	2012	EPIDOR (Basin Public Territorial Establishment)	2,400,000	1,200,000
Marais Audomarois	2013	Caps et Marais d'Opale Regional Nature Park and Saint-Omer Agglomeration Community	22,539	68,900
Mont Viso	2013 Cross-border since 2014	Queyras Regional Nature Park and Parco del Po Cuneese	427,080	292,369
Gorges du Gardon	2015	Syndicat Mixte des Gorges du Gardon	45,501	188,653

Table 1.3. *Primary characteristics of France's biosphere reserves*

1.4.3. *In Morocco*

Biosphere reserves seek to promote, outside of the network of protected areas, solutions consolidating biodiversity and its sustainable use (the global concept of land management) at the national level. Morocco has been officially committed from the beginning to a policy of creating the latest generation of biosphere reserves. This choice poses a number of problems with regard to implementation, as the concept and the governing model of biosphere reserves are still on the table in the institutional debate, 20 years after the declaration of the first biosphere reserve. The national master plan for protected areas from 1996 led to the creation of a national network of protected areas, which does not include biosphere reserves.

While they are still without recognition in any legal framework, biosphere reserves do form a part of national and regional strategies in the fight against desertification and the conservation of natural resources encouraged around the years from 1995 to 2000 (e.g. the Arganeraie and Oasis du Sud Marocain Biosphere Reserves) and of more recent strategies oriented around biological conservation encouraged since 2005 (the Intercontinental Biosphere of the Mediterranean and the Atlas Cedar Biosphere Reserve).

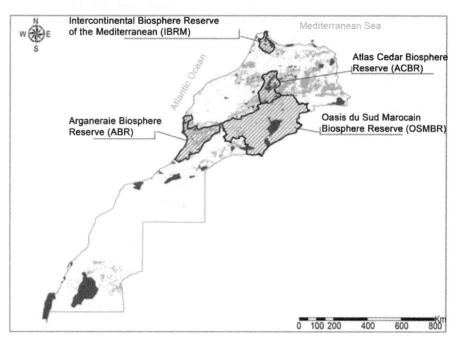

Figure 1.4. *Map of Morocco's Biosphere Reserves (and network of protected areas included in the biosphere reserves) (source: HCEFLCD 2016)*

The governance of each biosphere reserve differs, the Arganeraie and Intercontinental Biosphere Reserves being the most active today. They all follow an "authority model", governance in each case being top-down, and the managing body being dependent on a governmental authority at the regional level, of which regional water and forest departments are the catalysts.

Select committees are organized and coordinate multiple regional actors. The Regional Directorate of Water and Forestry and the Fight against Desertification (DREFLCD) of the High Commission for Water and Forestry and the Fight against Desertification (HCEFLCD, Ministry of Agriculture, Maritime Fisheries, Rural Development and Water and Forests) is primarily responsible for the conservation of nature and forests, the management of central areas, buffer zones and to a large degree transition areas: all forests are in the public domain, under the authority of HCEFLCD, both within and outside protected areas. Management can become difficult in areas not considered to be forests, where the HCEFLCD has no direct power of intervention. These non-forest areas come under the authority of various local administrations, and the managing body is often obliged to negotiate and coordinate with other institutions to implement decisions.

The Regional Directorate of Water and Forestry and the Fight against Desertification is recognized as the managing authority over biosphere reserves with regard to the following missions:

– the implementation of the biosphere reserve strategy;

– the application of the biosphere reserve development and management plan;

– the preparation of annual programs;

– the planning of activities;

– the preparation and signature of partnership agreements;

– administrative and financial management;

– compilation of activity reports;

– coordination of all involved actors.

Moreover, a network of local actors is equally involved in decision-making about the future of biosphere reserves: local/regional authorities (municipalities and "collectivités territoriales"), public administrations and institutions, professional sector organizations, universities and research institutions, NGOs and representatives of the participative wings of the national parks situated within the biosphere reserve. Its prerogatives are limited to the following missions:

– to examine and approve the biosphere reserve management plan submitted by the managing body;

– to formulate recommendations regarding any measures likely to improve the management of the biosphere reserve;

– to ensure coordination between the different local actors involved in the biosphere reserve and the consistency of interventions.

The participative body is presided over by the president of the regional council. An annual meeting is organized at the request of the participative body. However, the DREFLCD provides the secretariat of the participative body.

With regard to administrative and managerial responsibilities, the biosphere reserves are spread across two levels: I) the national level (the Parks and Natural Reserves Department of the HCEFLCD, in Rabat), which coordinates and supervises the national implementation of the UNESCO MAB program, all general matters concerning Moroccan biosphere reserves, and the governmental partnerships regarding them (e.g. cooperation with Germany on the Arganeraie Biosphere Reserve, cooperation with France on the Oasis du Sud Marocain and cooperation with Spain on the Intercontinental Biosphere Reserve of the Mediterranean); and II) the regional level, where each biosphere reserve deals with the mechanisms and area-specific policies of different governmental and administrative actors who are plotting their future.

1.4.4. *In Lebanon*

The Shouf and Jabal Moussa Biosphere Reserves are incorporated into the institutional landscape in a differentiated way.

The Shouf Biosphere Reserve (which is included in the Al Shouf Cedar Nature Reserve) is placed under the authority of the Lebanese Ministry of Environment, which runs it through the intermediary of the Committee for Protected Areas (APAC), responsible for the decision-making and the implementation of management objectives. The members of the APAC are volunteers, chosen by the Ministry, and representing local communities (municipalities, district commissions), independent environmental experts, environmental NGOs and institutional representatives (Ministry of Agriculture). The APAC ensures liaison with the biosphere reserve's team, which works on the day-to-day management and planning of the reserve.

Beyond the Ministry of Environment, other government agencies are involved: the Ministry for Public Works, responsible for the roads passing through the reserve; the Department of Antiquities (antiquities and ancient ruins); the Ministry of Information and the Maasser municipality (television broadcaster); and the Barouk Office of Water. The biosphere reserve needs the cooperation of all governmental and nongovernmental sectors to attain its objective, the conservation of natural and cultural heritage, with the improvement of well-being and income of the villages surrounding it.

The management team is responsible for the running of administration, management, technical activities, interview activities, protection, public awareness, communication and scientific research. It currently has 60 permanent employees and more than 40 temporary employees, who are all residents of the Shouf region.

The Association for the Protection of Jabal Moussa (APJM), a Lebanese nonprofit NGO, runs the Jabal Moussa Biosphere Reserve. It is the source of the accumulation of several international designation on top of that of UNESCO: from BirdLife International (Global IBA), from the UICN (as an example of private protected area "best practice"), from MedMAB, as well as the Lebanese protection labels of "protected forest" (from the Ministry of Agriculture) and "protected archeological site" (Ministry of Culture).

The APJM is composed of an assembly (which meets annually), an administrative council (which meets quarterly) and a management structure which oversees the day-to-day running of the reserve.

In 2012, the ALJM presented its first 10-year plan for sustainable management to the public (supported by the Swiss MAVA foundation and the UNESCO Regional Office in Beirut) during an interactive conference participated in by academics, researchers, local elected officials, religious representatives and local communities from the villages found within the reserve. This conference was the culmination of two years of work in cooperation with the above actors.

The primary management orientations in terms of biodiversity, culture and archaeology, academic research and above all of socio-economic development were defined on this occasion, and were made the object of a constant development and permanent dialogue between the parties involved in the 10-year plan from the beginning.

The 10-year plan for management and development should be updated in 2022.

1.5. Discussion

It is in Spain that the "UNESCO Biosphere Reserve" model has received the strongest and most consistent support: there, it occupies an important and recognized place, as much in legislature as within institutions. There are numerous high-profile sites, and the network that they make up is structured and well-organized.

The country strongly invests in international cooperation within the framework of the MAB program. It has organized three of the World Congresses that have structured the MAB program and the concept of the biosphere reserve, notably that of Seville in 1995, which produced the primary reference texts for the biosphere reserve, the Seville Strategy on Biosphere Reserves and the Statutory Framework for the World Network of Biosphere Reserves. It also supports the IberoMAB network (encompassing the Iberian Peninsula and Latin America) and more recently the Mediterranean Network of Biosphere Reserves.

In France, such a level of support is and continues to be lacking, which is certainly explained by the proximity of the concept of the biosphere reserve to the French concept of the Regional Nature Park, dating from 1970. There are 56 such Nature Parks (as of 2021), well structured within a federation, much like the biosphere reserves of Spain. Some of these, to increase their international profile, have also requested UNESCO designation as a biosphere reserve.

The originality of the French MAB program is tied to its permanent proximity to the world of research, which has led to the creation of an original structure, the MAB-France association, which is committed to reinforcing the links between the scientific world, ecology as a human and social science, and higher education, and the experimental spaces that are biosphere reserves, to promote an active transition towards increased sustainability. The force of this associative structure is the commitment of its members, a network of passionate, competent and engaged individuals.

This has resulted in innovative projects, but often with insufficient support, structure and circulation. Recent years have shown an improvement in its profile within the national landscape, as demonstrated by the recognition of biosphere reserves in law in 2016.

Adequate institutional positioning for the MAB program is clearly lacking in the two southern Mediterranean countries. However, it is notable that in all the countries studied, the integrated approach promoted from the beginning by the MAB program and formulated within the texts that underlie the biosphere reserve – the links between conservation, land management and development, the need for support through education and training, and understandings of relevant socio-economic dynamics – is a common thread through legislation and practice, as can be seen in the changes to the law surrounding protected areas in Morocco, and the 2006 law on French National Parks.

It is undeniable that the conceptual approach of the MAB and of biosphere reserves has spread, but has insufficiently informed the public policies of certain countries. Biosphere reserves are still thought of too often as mere international labels, with a lack of understanding of the concept's richness. It is not easy to give a detailed explanation of what a biosphere reserve is, and furthermore, they are harmed by their name, which seems rather inaccessible to the public at large.

These countries could, however, lean on the now mature experiences of others, of changes largely facilitated by the network.

The environmental urgency connected to the tardiness in taking action to limit crises of biodiversity and climate, despite repeated warnings from the scientific world for 50 or more years, and despite growing awareness in numerous countries, speaks (on a local as much as a global level) to an insufficiently productive dialogue in favor of the protection of fundamental resources. The MAB, in seeking to facilitate interaction between worlds that are too often kept separate, is more relevant than ever. The meeting of the worlds of higher education and of biosphere reserves, the mobilization of young generations, always highly motivated by this now 50-year-old program, is thus a winning venture, with regard to training, to engagement and citizenship, to the local management of land, to creating an international outlook, to cooperation and, above all, to the ability to tackle the profound changes that are surely to come.

1.6. References

Jardin, M. (2017). Man and the Biosphere Programme (MAB): Governance of biosphere reserves. In *International Worshop on MAB in South Caucasia*, 7–9 November, Turkey.

Jardin, M. (2020). 50 ans d'histoire du Programme MAB (l'Homme et la Biosphère) de l'Unesco. *MAB France* [Online]. Available at: https://www.mab-france.org/workspace/uploads/mab/documents/50-ans-histoire-du-mab.pdf.

Oñorbe, M. (2018). Las Reservas de la Biosfera españolas como modelos de gobernanza inspiradores de áreas protegidas. Bulletin 45, EUROPARC España [Online]. Available at: http://rerb.oapn.es/documentacion-y-difusion/publicaciones-y-documentos [Accessed 16 March 2020].

Organismo Autónomo de Parques Nacionales (2012). El programa MAB y su aplicación en España [Online]. Available at: http://rerb.oapn.es/images/PDF_publicaciones/programa_Mab_Espana_Sintesis.pdf [Accessed 27 April 2021].

Organismo Autonomo de Parques Nacionales (2019). Las Reservas de la Biosfera Españolas [Online]. Available at: http://rerb.oapn.es/images/PDF_publicaciones/AA_Informacion Basica_RRBB_Espa%C3%B1olas_2019.pdf [Accessed 27 April 2021].

Organismo Autónomo de Parques Nacionales (2021). Las Reservas de la Biosfera Españo-las – Mapa [Online]. Available at: http://rerb.oapn.es/red-espanola-de-reservas-de-la-biosfera/reservas-de-la-biosfera-espanolas/mapa [Accessed 27 April 2021].

Santamarina Arinas, R.J. (2015). Estudio comparativo de la situación jurídica de las RRBB españolas en la normativa oficial de cada comunidad autónoma en el que se destaque cuáles son las deficiencias más importantes en cada caso [Online]. Available at: http://rerb.oapn.es/images/PDF_publicaciones/Estudio_legislacion_Esp.pdf [Accessed 27 April 2021].

UNESCO (1996). Réserves de biosphère : la stratégie de Séville et le cadre statutaire du réseau mondial. Report, UNESCO, Paris.

UNESCO (2016). A new roadmap for the man and the biosphere (MAB) programme and its world network of biosphere reserves. MAB Strategy (2015–2025), Lima Action Plan (2016–2025), Lima Declaration. Report, UNESCO, Paris.

2

The Emergence and Evolution of Mediterranean Biosphere Reserves in France

The French Mediterranean region boasts the largest number of biosphere reserves in France, home to 7 of the 14 French sites. This sub-network is highly diverse, with humid zones (Camargue), scrublands (Gorges du Gardon) and mountains (from middling to high: Luberon Lure, Cévennes, Mont Ventoux and Falasorma Dui Sevi, the latter also notable for its insularity; the piedmont of Mont Viso is also Mediterranean).

It was also in the Mediterranean that the first biosphere reserves were created in 1977, which have since had to change their perimeters, governance and methods with the arrival of the Statutory Framework of the World Network of Biosphere Reserves. These reserves may thus also speak to the development of ideas within the MAB program, and within the world of biosphere reserves.

2.1. Profound changes across first-generation sites (1977)

The Camargue Biosphere Reserve corresponded to the perimeter of the Camargue Nature Reserve, run by the SNPN (National Society for the Protection of Nature). It included the Étang de Vaccarès. Uninhabited, it nonetheless bore witness to a strong interaction between humans and nature, the current humid areas around the Rhône Delta (with their powerful ecological potential) being the result of ancient interactions with humans (embankment, modifications allowing for the management of water levels and salinity, human activities such as fishing, the rearing of

---

Chapter written by Catherine CIBIEN.

livestock, the agriculture of rice, grapes, grains, etc.). Beginning with the adoption of the Statutory Framework in 1995, it became evident that this biosphere reserve did not meet the required criteria and would need to be enlarged, the initial site, already granted the status of a national natural reserve, possibly becoming a core area in the future biosphere reserve. This necessitated the construction of a territorial project around this nature reserve. The existence of a regional nature park on the eastern part of the Rhône, and of other areas protected by other apparatuses (coastal conservatories, voluntary natural reserves, etc.), could not but involve new actors and the adaptation of the reserve's governance.

Figure 2.1. *Landscape of the Camargue Biosphere Reserve (source: PNR Camargue)*

MAB France needed to act as a mediator, inviting all actors across the Rhône Delta (elected officials, administrations representing socio-professionals from within agriculture, fishing, livestock rearing, tourism, the reed industry, protected areas, cultural associations, etc.) to neutral ground, to discuss either requesting that UNESCO withdraws Camargue's designation as a biosphere reserve, or allowing the actors to organize between themselves a proposal for a biosphere reserve encompassing the extended Camargue area, one responding to the criteria put in place by the Statutory Framework. The group chose the second option and the application for a fresh designation for the Camargue Biosphere Reserve, extended across the entire Rhône Delta, was approved by UNESCO in 2006. Straddling two regions (Provence-Alpes-Côte d'Azur and Occitanie) and two departments, the management of its east side is the responsibility of the Camargue Regional Natural

Park, and the west side is the responsibility of the Camargue Gardoise Syndicat Mixte. It is endowed with a scientific council, and its management documents are those of the bodies in place attached to the development of projects (conservatory, touristic, etc.) at the level of the delta as a whole, the geographical, ecological and cultural cohesion of which is undeniable.

Figure 2.2. *Landscape of the Fango Valley (source: J. Innocenzi)*

The Fango Biosphere Reserve was recognized by UNESCO in 1977 after the submission of a short proposal written by scientists from the APEEM, demonstrating the uniqueness of the area's groves of Holm oak. At the time, the proposal made no reference to governance or any management policy. The biosphere reserve would be extended for the first time in 1990, expanding to cover the entire drainage basin of the Fango river. The "Fango MAB Committee" was created, gathering the APEEM, the ONF, which managed a good deal of the forest, and the Corsica Regional Nature Park to discuss its management. It was paired with a management committee and a scientific council. The Fango Valley has a strong geographical coherence, undoubted ecological, cultural and historical interest, but with its hundreds of very rural inhabitants, it could only with difficulty be seen as a model for land management. At the time of its periodical review, it was thus decided that the reserve should be extended again to meet UNESCO's recommendations. The regional nature park was made responsible. The extension process was tied to an updating of the regional nature park's charter and synergies between the bodies were sought, the charter becoming the primary management document: it was decided in the end that the biosphere reserve would be extended to cover the maritime aspect of the regional nature park and would notably include the coastal site listed as a UNESCO World Heritage Site. Recognized by UNESCO in 2020, the reserve changed its name (Falasorma-Dui Sevi) but kept its scientific council and management committee, which created for it a specific management plan.

2.2. The recognition of local development projects promoting natural and cultural heritage

The Cévennes Biosphere Reserve was created in 1985 to promote the idea, once unpopular, of an inhabited national park with a strong cultural component, the result of ancient interactions between humans and nature. The interaction between conservation and development and the integrated vision were affirmed by this, despite a 1960 law on national parks lending itself poorly to the idea. With its overhaul in 2006, the compatibility of the foundational texts of the MAB program with those of the new national parks would be improved: the perimeter, the zoning, the authorities and the management policies were made coherent, and began to overlap.

Figure 2.3. *High-altitude landscapes in the Cévennes Biosphere Reserve (source: J.-P. Malafosse)*

Figure 2.4. *Lavender fields in the Luberon-Lure Biosphere Reserve (source: PNR Luberon)*

Figure 2.5. *Aerial view of the Gorges du Gardon (source: SMGG)*

Luberon is one of the oldest regional nature parks in France, renowned for the quality of its management. The park wished to reach for new heights and international recognition through being named as a biosphere reserve: an active and structured educational service was thus established, as well as a scientific council. At its first periodical review, the biosphere reserve framework appeared to be subtle, and well adapted to the cooperation needed to develop the work habits and projects with the authorities situated to the east of the park (on Lure Mountain) that had been envisaged. It was thus extended to the domains of these authorities, and its name was changed, becoming the Luberon-Lure Biosphere Reserve.

The Gorges du Gardon, home to the Pont du Gard, listed as a UNESCO World Heritage Site, is an exceptional site at the heart of the scrubland stretching from Nîmes to Uzès. This region, incredibly biodiverse and home to important cultural heritage, had nevertheless been relegated to the shadow of Camargue and Cévennes. No wonder it hoped to achieve its own recognition through becoming a biosphere reserve in 2014!

The transfrontier Mont Viso Biosphere Reserve is above all mountainous, peaking at 3,841 meters on the Italian side. Situated at the edge of the Mediterranean basin, the zone's uplands are marked nevertheless by Southern French vegetation, and is attached to the Provence-Alpes-Côte d'Azur at the administrative level. On the French side, the reserve takes in the Queyras Regional Nature Park, which supports its running, and the communities of the municipalities surrounding it. The biosphere reserve aims to strengthen the cooperation between these bodies and the Italian half that has existed historically.

Figure 2.6. *Environmental education in the Mont Viso Biosphere Reserve (source: Christophe Gerrer)*

2.3. References

Cibien, C. (2006). Actualités de la recherche Les Réserves de biosphère : des lieux de collaboration entre chercheurs et gestionnaires en faveur de la biodiversité. *Natures sciences sociétés*, 14, 84–90.

Hervé, C., Jacob, T., Sagna, R., Cibien, C. (2022). Identifier les activités scientifiques dans les réserves de biosphère françaises : une chasse au trésor ? *Natures sciences sociétés*, 30(2022/1), 3–13.

3

Perspectives on Mediterranean Biosphere Reserves

3.1. Close-up on the strengthening of the Mont Ventoux Biosphere Reserve's governance

3.1.1. *Introduction*

Country	France
Name of the biosphere reserve	Mont Ventoux
Date of UNESCO designation	1990
Management structure	Mont Ventoux Regional Nature Park
Area	90,000 ha
Population	83,500 ha

Table 3.1. *Profile of the Mont Ventoux Biosphere Reserve*

3.1.2. *An iconic Mediterranean mountain*

As a veritable microcosm of the Mediterranean Alps, the Mont Ventoux Biosphere Reserve is organized around the "Giant of Provence", which is characterized by an impressive biological stratification. Mediterranean and alpine habitats and species are juxtaposed, thanks to, on the one hand, a steep gradient with regard to altitude (reaching 1,909 meters at the summit), and strong contrasts between the southern and northern slopes, on the other hand.

Chapter written by Ken REYNA, Martí BOADA and Mchich DERAK.

In his *Souvenirs Entomologiques*, Jean-Henri Fabre (1879) thus recalls:

> A half-day's vertical movement allows to pass before one a succession of the principal forms of vegetation that one is likely to encounter during a long journey from north to south, following the same meridian. On departing, one's feet tread soothing tufts of thyme, which form an uninterrupted carpet across the lower summits; in some hours, they will tread the dark cushions of opposite-leaved golden saxifrage, the first plant to offer itself up, in June, to the botanist setting out upon the Spitzbergen shores.

Mont Ventoux is a land of contrasts, rendered so by its altitude and its geographical situation as an intermediary space between the temperate and Mediterranean spheres. The mountain thus offers a stunning range of natural environments, home to countless plant and animal species, some of which are considered to be genuine rarities.

3.1.3. Conserving and developing the assets of an exceptional area

The range of altitudes, environments and climatic conditions throughout Ventoux have allowed for a flora and fauna which are both rich and remarkable: 150 species of nesting birds, 20 species of bats, 14 of reptiles, 8 of amphibians, 2,500 of insects (1,425 species of butterflies), countless species of fish (upper reaches of river rich in fish), etc. Thanks to its bioclimatic tiering, 1,500 species of plants have been able to evolve, some of them, such as the *acis fabrei* (*nivéole de Fabre* in French), being strictly endemic.

Ventoux is also the site of ancient human activity, responsible for profound transformations: pastoralism, and its Mediterranean accompaniment, fire, appear to be at the root of the first deforestations from the Neolithic period, a practice intensified between 5,000 and 4,000 BCE, being pursued as part of Gallo-Roman exploitation of forests and through the Middle Ages, up until the 19th century. Large-scale reforesting projects were carried out at the end of the 19th century, and the forest expanded, replanted with local species (green or pubescent oak, for example) and non-native species (Austrian pine and Atlas cedar in particular). The local economy evolved towards grapes, lavender and grains, and then towards arboriculture, market gardening and tourism.

3.1.4. Governance evolving with the times

The designation of the Mont Ventoux Biosphere Reserve by UNESCO in 1990 was taken as a chance to initiate a policy for the protection and management of natural spaces and biodiversity. Different tools for protection and management were gradually deployed:

- an integral national biological reserve (906 ha);
- six prefectural decrees for biotope protection (2.126 ha);
- three Natura 2000 special conservation zones (5,620 ha);
- six sensitive natural areas (597 ha).

It is a local authority – the Mont Ventoux Management and Equipment Syndicat Mixte – which is at this time in charge of the running of the biosphere reserve. It is equipped with a management committee made up of elected officials, social and economic actors, members of environmental and cultural associations, etc. The committee's role is advisory. It is a space for discussing the management of the massif, the challenges of the territory, etc. The Syndicat Mixte, made up exclusively of elected officials, maintains the decision-making power (voting on budgets, plans of action, etc.).

At the time of its creation, this institutional blend was very modern, the majority of biosphere reserves being attached, rather, to existing protected areas: in the case of national parks, this combination gave them functions for support and development that had previously been mostly absent; in that of regional nature parks, it gave them an international and scientific dimension that until then had remained slight.

The Mont Ventoux Biosphere Reserve would go on to lead a host of projects, in the domains of biodiversity and heritage (renewing understandings, tracking programs, management projects, etc.), of ecodevelopment (organizing cycling tourism, developing nature sports, deploying eco-actors, a country bistro network), of education, etc.

However, it is clear that there is room for improvement both in terms of the human and financial resources at the disposal of the biosphere reserve and in terms of its actual interventions. While the ambitions and functions of a regional nature park (RNP) and a biosphere reserve can be considered to be approximately the same, the major difference resides in the fact that the RNP is an apparatus of French law, while the biosphere reserve has no such attachment, and is considered rather an international label. It can thus only with great difficulty obtain regular funding with which to carry out its functions, and the legitimacy of its interventions or declarations can easily be questioned.

To alleviate these difficulties – the value of an integrated approach to conservation and development being recognized locally – a proposal for an RNP was developed. It would take nearly 10 years to establish the park, and since its establishment, the RNP has become the support for the Mont Ventoux Biosphere Reserve. It presents the following advantages:

– a management policy (the RNP charter) with which all relevant authorities are officially engaged for 15 years;

– a strong state accompaniment through the establishment of the RNP;

– funding already earmarked for RNPs.

The Mont Ventoux Biosphere Reserve has since relied on the strong institutional and legal apparatus that is the regional nature park. Its anteriority, and the land debates and work methods that the biosphere reserve had already introduced, were considerable assets in the construction of the RNP. The complementarity of the two apparatuses, their relationship, the reinforcement of interventional legitimacy and the stronger involvement of authorities are all strong contributions to the biosphere reserve's cause.

The action plans also found themselves strongly supported, as well as the human resources allocated, with regard to regulation (the natural reserve project, studies for the implementation of a classed site around Ventoux, the extension of the prefectoral decrees for the protection of the biotope), the management of natural spaces (new sensitive natural spaces, the revision of Natura 2000 management plans, the employment of new scientific tracking of natural habitats and species: high-altitude fields, the Little owl, the Egyptian vulture, etc.), public awareness (the launching of municipal biodiversity atlases, etc.) and economic development (short supply circuits, the integrated management of nature sports: trail running, mountain biking, cycling, hiking, climbing, etc.).

3.2. Close-up on the Montseny Biosphere Reserve

Country	Spain
Name of the biosphere reserve	Montseny
Date of UNESCO designation	1978
Management structure	Sustainable area of the Barcelona Provincial Council, and environmental area of the Girone Provincial Council
Area	50,166 ha
Population	51,573 ha (2014)

Table 3.2. *Profile of the Montseny Biosphere Reserve*

The Montseny Biosphere Reserve boasts one of the highest levels of biodiversity in the entire Mediterranean basin, straddling three bio-geographic regions, a blend of the different landscapes of Western Europe, a refuge for numerous taxons. Highly sensitive to the global changes presently occurring, it is a highly popular area for scientific research. Its cultural and historical values make its importance universal.

Figure 3.1. *Monsteny in autumn (source: M. Boada)*

3.3. Close-up on the Menorca Biosphere Reserve

Country	Spain
Name of the biosphere reserve	Menorca
Date of UNESCO designation	1993
Management structure	Menorca Biosphere Reserve Agency, within the Menorca Insular Council
Area	514,485 ha
Population	95,000 ha

Table 3.3. *Profile of the Menorca Biosphere Reserve*

Menorca (Minorca) is the easternmost island of the Balearics. The biosphere reserve encompasses the entire nonmarine surface of the island, and a large marine area, including the nature park of S'Albufera des Grau. All of the municipalities of this humanized land, comprised of traditional rural landscapes, are engaged with the biosphere reserve. The diversity of Mediterranean habitats is remarkable: ravines, grottos, trenches, humid zones, dunes, islets, small hills and hillocks, broad and rural open spaces, etc. The archeological and historical heritage of the reserve – more than 1,500 archeological deposits – is exceptional.

Figure 3.2. *Naveta des Tudons (source: S. Sanchez)*

3.4. Close-up on environmental education and SDGs, an opportunity for Mediterranean Biosphere Reserves

It is clear that we find ourselves in a historically unprecedented situation of *civilizational crisis* (universal), which is to say, not one of a *crisis of civilization* (local). The manifestations of this crisis are numerous and express themselves in a variety of ways.

All of this unfolds in a context in which it is understood that the process is accelerating, which is reducing the ability of society to react with sufficient speed and urgency.

The UN's declaration of a climate emergency is the strongest evidence for the gravity of this historical moment, a moment aggravated by the Covid-19 pandemic.

The institutional tools put in place around the 2030 Agenda and its 17 sustainable development goals (SDGs) can be seen as a promising start, and biosphere reserves (BRs) as an appropriate terrain for experimentation.

The question is to understand whether or not there is sufficient time left to correct the environmental crisis which has appeared. We hope that such a correction is the path we will follow, but we cannot wait another minute. As Bellamy reminds us, those among us who are involved in the management and diffusion of sustainability cannot be content to dawdle along when the forest is burning down around us. Contradictory disciplinary positions, corporate discussions and exclusive truths help no one; we all need one another at this moment of crisis.

From an educational perspective, and notably to the end of raising awareness of the situation and of the necessary engagement with sustainable development objectives, it seems strongly advisable to go beyond the clichés. As we all know, clichés are as far as we can get from a true understanding.

It is advisable to train others, to help them understand (the IUCN's One Earth document). We should not scare the pupils, students and adults at whom our message is aimed through extreme alarmism. As Paul Ehrlich points out, if we scare our audience, through simple defense mechanisms, they will turn their backs on the problem. We need to remind ourselves, and indeed train ourselves, to understand, because, as Margalef suggests, well-informed people multiply where poorly informed people dwindle.

The situation is without doubt urgent, but as a Spanish aristocrat once said to her valet: "Dress me slowly, I'm in a hurry"[1].

And if biosphere reserves are privileged spaces for experimentation in the fields of conservation and biodiversity, they are also privileged spaces for experimentation in environmental education. In this sense, we wish to recall that biodiversity, as well as its cultural and social dimensions, are extremely stimulating notions, as much for pupils as for adult society. A well-preserved landscape is above all an exceptional classroom.

Socio-ecosystems are themselves a university, where learning opportunities are limitless. It is very strongly advisable to promote environmental understanding and to aid biosphere reserves which, in our case with its different variables, constitute cultural and ecological microcosms of the Mediterranean, with which term represents in terms of biogeography and civilization in its entirety.

Faced with accelerating climate change, BRs are becoming bellwether areas, early indicators of changes to biomes stemming from modifications in the climactic environment. In this context, biodiversity, on top of its undeniable value as heritage which must certainly be protected, becomes a superlative indicator of changes produced within lands and landscapes because of severe environmental alterations. In these conditions, beyond the values and curiosities of the biodiversity of BRs, by applying appropriate models, they can represent a highly rigorous indicator of sustainability.

---

1 Note from the author: we have no ideological sympathy with the aristocrat; this is simply a metaphor in service of a pedagogical reflection which implies a real urgency.

In their turn, BRs can act as a space for creating relationships of cooperation and shared understandings between the academic world, the research derived from that world, and the citizenry. In this meeting space, administrators play a key role. They must put forward strategies for overcoming the conflicts between formal and informal education, and work on participative processes with populations linked to BRs, to attain, as far possible, the sustainable development objectives.

3.5. Close-up on the Intercontinental Biosphere Reserve of the Mediterranean[2]

Country	Spain
	Morocco
Name of the biosphere reserve	Intercontinental Biosphere Reserve of the Mediterranean
Date of UNESCO designation	2006
Management structure	RBIM Joint Committee (Moroccan Department of Water and Forestry and the Spanish Ministry for Ecological Transition, and the Andalusian Regional Government)
Area	907.185 ha including 81.436 ha of central zones
	Morocco: 65.342 ha
	Spain: 16,094 ha
	Marine zone between countries: 13,050 ha
Population	529.086 ha including 402,227 ha in Morocco and 126,589 ha in Spain
Protected areas	Morocco: 1 national park, 1 future national park, 5 SIBEs
	Spain: 4 nature parks, 4 natural areas, 5 natural monuments, 13 ZECs, 10 ZEPA declarations

Table 3.4. *Profile of the Intercontinental Biosphere Reserve of the Mediterranean*

The Intercontinental Biosphere Reserve of the Mediterranean is the first intercontinental biosphere reserve to be designated by UNESCO. It contains a marine corridor and aims towards communication and collaboration between the two continents with their different conditions: Europe and Africa.

Between the south of Andalusia and the north of Morocco, it forms an open arc towards the Mediterranean, split by the Strait of Gibraltar. The uniqueness of this natural and sociocultural crossroads, where the influence of the Atlantic is decisive,

---

2 Based on: www.unesco.org/mab.

is illustrated by its primary axes (mountain, water, culture) and by the environmental and socio-economic complementarity between its northern and southern parts.

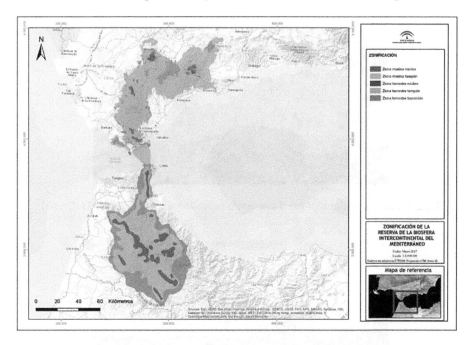

Figure 3.3. *Zoning of the Intercontinental Biosphere Reserve of the Mediterranean. Moroccan Department of Water and Forestry (2017). For a color version of this figure, see www.iste.co.uk/romagny/biosphere2.zip*

The natural landscape of the biosphere reserve is comprised of the mountains of the large and rugged Baetic System, which stretches across Andalusia and northern Morocco, ranging vertically from 2,170 meters at Jbel Lakraa to more than 500 meters in the strait of Gibraltar, thus creating natural conditions shared across the opposing shores. These mountains, with their abrupt reliefs, are rich in water. They are home to more than 2,000 species of flora and fauna, and offer some important endemic species: the Moroccan fir, the Spanish fir, the black juniper and the Atlas cedar. The Strait of Gibraltar is a major site for bird migration (more than 1,000 birds passing through every hour), the passage of different cetaceans and biotic construction through coral.

Figure 3.4. *Landscape of the Tallasemtane National Park (source: M. Derak)*

Figure 3.5. *Humanized landscapes of Bouhachem (source: M. Derak)*

The biosphere reserve is an attractive area for visitors, thanks to the beauty of its landscapes combined with a socio-economic system typical of mountainous areas near to the sea. Traditional uses of natural resources are combined in such areas with agriculture, livestock rearing, fishing and forestry, with communication and commercialization being made possible through their ports. With the promotion of tourism, the administrations on both sides of the Strait hope that those who visit this space will be able to observe the consistency of natural, historical and social values across the Intercontinental Biosphere Reserve.

The Euro-African frontier zone situated between the state of Andalusia and the Rif thus shares a historical, politico-economic, civilizational and cultural basis, an orographic and environmental resemblance and a biogeographic continuity.

Figure 3.6. *A street in Chefchaouen (source: M. Derak)*

The most remarkable characteristic of this biosphere reserve is its wish to promote a sustainable development model within a framework of institutional collaboration. This wish is manifested in the action plan of the Intercontinental Biosphere Reserve which works to promote sustainable development, the improvement of environmental conditions, and the strengthening of the reserve and its management.

Thus, so as to enhance and to better conserve its natural and cultural resources, the reserve is engaged in diverse projects contributing to:

– the promotion of tools with which to overcome the discontinuities of natural systems, and to work towards cooperation between local societies;

– the promotion of shared historical, cultural and natural features;

– integrated development (agricultural, rural, environmental, cultural and touristic) on both halves of the reserve;

– the promotion of scientific and technical exchange and the training of administrators for protected areas.

The major challenge of the Intercontinental Biosphere Reserve is the search for a compromise between the continuation of traditional activities, still alive on both sides of the Mediterranean, the development of new activities such as tourism, and the modernization of modes of production with regard to agriculture, livestock rearing and industrial production.

3.6. References

Cibien, C. (2006). Actualités de la recherche Les Réserves de biosphère : des lieux de collaboration entre chercheurs et gestionnaires en faveur de la biodiversité. *Natures sciences sociétés*, 14, 84–90.

Flamme, G., Maneja, R., Marlès, J., Bontempi, A., Jesus Gracia, J., Corbera Serrajordia, M., Floguera, G. (eds) (2019). *Réseau des réserves de biosphère Méditerranéennes*. Fondación Abertis, Castellet.

Hervé, C., Jacob, T., Sagna, R., Cibien, C. (2022). Identifier les activités scientifiques dans les réserves de biosphère françaises : une chasse au trésor ? *Natures sciences sociétés*, 30(2022/1), 3–13.

4

From the Ecological Quality Status Evaluation to the Knowledge Transferability. A Cross-cutting Experience in Montseny Biosphere Reserve

4.1. Introduction

Since the launch in 1971 of the Man and Biosphere (MAB) program, UNESCO promotes biosphere reserves worldwide as learning sites, living laboratories for the understanding of interactions between humans and their environments. Among their functions, the logistic support through research, monitoring, education and training plays a role of key relevance. The case of Observatori Rivus environmental monitoring and education program is an exemplary case in point, as it embraces the MAB goals, embodies them in an innovative monitoring-education methodology and has applied it in the Montseny Biosphere Reserve since 1996.

The long-term monitoring of the socio-ecological indicators of the quality status of the river basins inside the reserve informs and underpins an environmental education program (PROECA) for the knowledge transfer to society under the motto "to bring back citizens to the river".

Chapter written by Sònia SÀNCHEZ-MATEO, Antoni MAS-PONCE and Roser MANEJA.

4.2. Mediterranean river basins as valuable and complex socio-ecosystems

Mediterranean river basins are mainly characterized by their inter-annual climatic variability. Drought periods in summer alternate to mild winters (Cuddenec et al. 2007). Precipitations in these areas are linked to the seasonality, mostly in spring and autumn when flood episodes are more remarkable. This climatic variability will be more intensified due to Climate Change, the average annual temperatures on these areas are now approximately 1.5°C higher than the preindustrial period (1880–1899) and may be warmer in 2040 by 2.2°C (MedECC 2019). Freshwater resources in the Mediterranean basin, considering a 2°C warming, will decrease by 2–15%, which is the largest freshwater decrease in the world (Gudmundsson et al. 2016). In the specific case of Catalonia (north-east Spain), climatic models also indicate an upward temperature trend during the coming decades. From the period 1971–2000 to 2031–2050, models indicate that the annual average temperature in Catalonia will increase by 1.4°C. Moreover, projections also show an annual rainfall decrease of 7% (CADS et al. 2017).

Historically, some of the Mediterranean fluvial systems became the core element of the society and the territory around them where most human settlements and infrastructures were strongly established during the 1970s and 1980s. Due to the intensification of these human activities, mainly due to the increase of the population density, the Mediterranean region has suffered pronounced negative pressures and impacts on its environment (Blondel et al. 2010). Moreover, during the fast industrialization process of these areas, profound economic changes which took place in this region led to a process of land abandonment in rural areas, especially in headwaters (Otero et al. 2011; Sànchez-Mateo et al. 2011; Voza et al. 2015). These processes have negative effects on the freshwater ecological quality and the flow regime as consequences (Vega et al. 1998; Hermoso et al. 2011).

4.2.1. The evaluation of ecological quality status

Water is an essential part of life; it is more than a mere resource. It has an ecological as well as a social function, so its protection and conservation is essential. The Water Framework Directive 2000/60/EC (hereinafter WFD) is the European regulatory framework that lays the groundwork for water management with a new focus on aquatic ecosystems as well as participation, incorporating citizens into decision-making.

WFD also establishes common bases for the integrated diagnosis of the quality status of water bodies from a physicochemical point of view, as well as from a hydromorphological and biological perspective. Furthermore, it urges the attainment of a good quality status.

The ecological status is the measure of quality of the functionality and the structure of an aquatic ecosystem. Moreover, a good ecological status is achieved when the biological communities are equal or close to those that are under unchanged conditions, and hydromorphological and physicochemical conditions should allow for the proper development of these communities.

Climate, orography, soil, vegetation and human activities determine the aquatic ecosystems characteristics: hydrology, chemical composition and water temperature, light gradient, habitat structure, biotic interactions (competence, depredation, etc.) and trophic resources availability. Aquatic organisms and those directly linked to fluvial systems are subjected to all of these factors and so, they provide ecosystems' status information. In this sense, most of all of the aquatic organisms can be used as fluvial systems indicators applying tested protocols and methodologies that permit mid- and long-term monitoring to assess quality status.

4.2.2. *Knowledge transfer and environmental education*

Environmental education is considered to be an essential tool under the current context of global environmental crisis. In the last decades, since the second half of the 20th century and especially in the northeastern area of the Mediterranean basin, the increase of human population density in urban areas, the settlement of industrial zones mostly concentrated along fluvial courses (Duran 2018) and the construction of infrastructures with repercussions on the territory, among other factors, have led to a process of disconnection between society and their environment. This progression has resulted in a general loss of knowledge and awareness with respect to the territory. In this context, environmental education strategies can contribute to reverse this process.

It was not until the First World Conference on the Environment organized by the United Nations in Stockholm (1972) that the Environmental Education and Communication concept was finally defined at an international level. This conference was celebrated in order to discuss "The limits of Growth" elaborated by the Rome Club in 1972, which enhances the relevance of the negative effects of the current development model and with the main goal of finding solutions to solve the environmental crisis.

The environment as a pedagogical resource was starting to be used in Europe at the end of the 1960s and 1970s with the need to promote a social movement to face the general perception of the environmental crisis and to motivate an active education that promotes the direct contact with nature.

Some researchers highlighted the importance of the experiences with the nature since for childhood to promote social and ecological awareness to face the distresses that alter natural dynamics and to increase the socio-ecological knowledge with respect to the environment (Corraliza and Collado 2013).

Observatori Rivus, a multidisciplinary applied research project led by Fundació Rivus and Autonomous University of Barcelona, aims to contribute to a reconnection between society and fluvial systems by the implementation of a knowledge transfer strategy through an environmental education, communication and training program, called PROECA.

4.3. Study area: Montseny Biosphere Reserve

Located in the Catalan Pre-coastal (mountain) Range, in the NE Iberian Peninsula, Montseny was declared to be a UNESCO Biosphere Reserve in 1978 under the purpose of achieving the necessary balance and synergies between society and the conservation of natural and cultural values.

It has a total surface of 50,166.63 ha divided into the core zone (9,058.07 ha, 18%), the buffer zone (22,914.00 ha, 45.7%) and the transition zone (18,194.56 ha, 36.3%). The highest altitude is reached at the Turó de l'Home peak, at 1,707 m, followed by Agudes (1,706 m) and Matagalls (1,693 m). The biosphere reserve comprises 18 municipalities with a total of 51,573 inhabitants (Figure 4.2). The geographical location of Montseny and the presence of three of the main biogeographical regions of Western Europe (Mediterranean, Euro-Siberian and Boreoalpine) confer to this area the range limit of different organisms and ecosystems, where northern and southern distributional boundaries for different species converge. This biogeographical uniqueness is exposed to global change and especially to climate change, becoming a very sensitive area in front of these processes and so considered as a sentinel landscape where applied and experimental research acquire remarkable relief.

Therefore, the Montseny Biosphere Reserve is one of the most emblematic natural areas in the region, where typical elements from a humanized landscape stemming from ancient times fuse with biogeographical values (Figure 4.3). Together with its proximity to the urban and metropolitan region of Barcelona, the ecological and cultural value confers to Montseny a great popularity that attracts a large number of visitors, being the service sector and touristic activities – above the primary sector, mainly related to livestock and forestry – essential for the socio-economic of the region. Throughout the twentieth century, Montseny went through a significant transformation process due to energy changes associated with industrialization, urban growth and a decline in the primary sector. Currently, the Montseny Biosphere Reserve also plays a notable role in the provision of environmental services and biodiversity conservation as well as in cultural and pedagogical activities linked to socioenvironmental awareness.

4.3.1. Observatori Rivus, a cross-cutting project in Mediterranean river basins

Observatori Rivus is an inter- and multidisciplinary research project that aims to establish a set of indicators to carry out the evaluation of quality status through medium- and long-term monitoring in Mediterranean fluvial systems, specifically in Besòs and Tordera river basins (NE Catalonia).

Thus, the research focuses on the monitoring and long-term registration of biological, hydromorphological and physicochemical indicators performed by different research areas where methodological protocols are developed and tested (Figure 4.1).

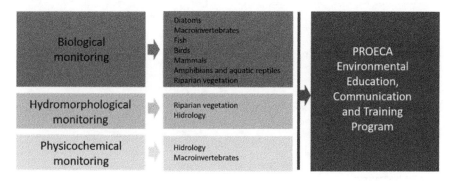

Figure 4.1. *Organizational structure of Observatori Rivus, highlighting PROECA as the key tool for knowledge transfer and dissemination. For a color version of this figure, see www.iste.co.uk/romagny/biosphere2.zip*

The long time series data obtained by Observatori Rivus project, started in 1996, facilitate the systematic assessment of the quality status of the fluvial system, essential for detecting changes in the period studied.

Moreover, the project develops and promotes knowledge transfer activities thanks to its own environmental education, communication and training program (PROECA), launched for the first time in 2004. The main aim of this program is to communicate the research results obtained to the general public, academics and policy makers alike. On the other hand, PROECA fosters a contribution to improve the relationship between fluvial systems and society. Connections to enhance this engagement are established by means of participative initiatives, approaching people to the river and involving participants in the application of methodologies for quality indicators assessment.

4.3.2. Sampling units

Observatori Rivus develops its tasks in the Besòs and la Tordera river basins (NE Catalonia), including part of the Montseny Biosphere Reserve considering headwaters and mid-courses of la Tordera, Gualba, Breda and Arbúcies streams, all of them belonging to la Tordera river basin; and Vallforners, l'Avencó and Congost streams in the Besòs river basin (Figures 4.2 and 4.3).

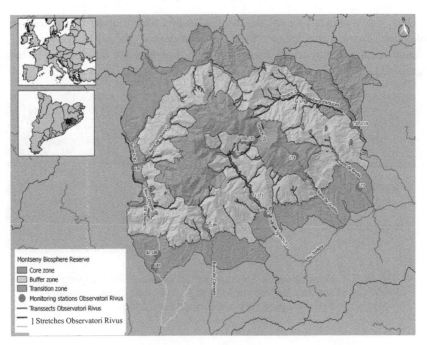

Figure 4.2. *Map of the Montseny Biosphere Reserve displaying the sampling units of Observatori Rivus. For a color version of this figure, see www.iste.co.uk/romagny/biosphere2.zip*

Figure 4.3. *Riparian areas in the Montseny Biosphere Reserve. From left to right and from top to bottom: headwaters of la Tordera in La Castanya valley; headwaters of Gualba stream in Santa Fe valley; two images of la Tordera in headwaters areas dominated by riparian forests with alder (Alnus glutinosa) and common ash (Fraxinus excelsior); la Tordera in the mid-course; and Arbúcies stream*

Three sampling units are determined to comply with the methodological protocols related to the above-mentioned monitoring areas:

– Stretches: defined by the sections in which a river course is divided under a criterion of homogeneity in terms of their physical characteristics and their degree of naturalness or disturbance. Most stretches are coincident with water body masses defined by the Catalan Water Agency. These sections contain the other two types of sampling units: transects and stations, both of them used as basic monitoring units.

– Transects: defined as longitudinal sections along the river being representative of the stretch where they are included. Their longitude is between 500 meters and 3 kilometers, in accordance to the requirements for each of the research area protocols and methodologies. Avifauna, riparian vegetation (quality of the riparian forest), mammals and amphibians and aquatic reptiles are monitored in transects.

– Stations: defined as specific and punctual sampling units suitable for the methodological application of macroinvertebrates, diatoms, fishes, hydrology, amphibians and aquatic reptiles (HDP: high-diversity ponds) and riparian vegetation (flora inventories) research areas.

4.4. Research areas

Observatori Rivus project focuses on eight research areas for biological, hydrological and physicochemical monitoring. Each of them follows a specific methodology based on validated protocols and different campaigns are performed throughout the year in order to evaluate intra-annual variability in relation to the abundance and composition of communities.

4.4.1. Biological monitoring

4.4.1.1. Diatoms

Benthic diatoms, given their microscopic size and high reproduction rate, respond sensitively and quickly to changes in their environment, so that they reflect the impacts that occur at the micro- and mesoscale. Therefore, these microscopic algae are very good bioindicators of variations in the physicochemical quality status of water and the biological quality of rivers. Due to their microscopic nature, alterations in the physical characteristics of the river (variations in water flow and velocity or changes in the granulometry of the riverbed) affect them to a lesser extent than in other organisms such as macroinvertebrates.

Through the evaluation of diatoms, the aim of this research area is to determine over time and space the quality of water in river courses from the specific diversity of benthic diatom communities using the IPS index (Gomà 2005; ACA 2006).

4.4.1.2. Macroinvertebrates

The benthic macroinvertebrate community is historically one of the most widely used bioindicators to determine the biological quality of river and stream water (Rosenberg and Resh 1993), and it is currently one of the quality elements that are part of monitoring and control programs in the whole European community, following the guidelines of the WFD. Sampling protocols, biological indices and their quality ranges according to each type of river are defined by the Catalan Water Agency (ACA 2006).

The methodology consists of the collection of biological samples of macroinvertebrates from the sampling units established for the determination of the biological quality of water with biological indices (IBMWP, BMWPC and FBILL). The fluvial habitat index (IHF) is applied complementarily to evaluate the general conditions of the complexity of the habitat in Mediterranean rivers, closely related to the biological indices.

4.4.1.3. Fish

Fish tend to occupy high trophic levels (Sánchez-Hernández and Amundsen 2015) and are relatively easy to sample and to identify. In general terms, their absence indicates limitations in the ecological status in the study area (Lorenz et al. 2013), so this group has often been used as indicator of the health for the aquatic ecosystems. In addition, they are well known by the general public and become an excellent communication and environmental management tool (Simon 1999; Benejam et al. 2008a; García-Berthou et al. 2016).

The use of biotic water quality indices based on the composition of fish communities is a common practice, so the WFD has selected fish, together with macroinvertebrates and diatoms, as indicators of ecological status in relation to biological monitoring.

The fish research area Observatori Rivus uses the IBICAT2b index (García-Berthou et al. 2016) to determine the ecological status of the sampling units. In addition, the analysis of the composition and amount of species captured during sampling makes it possible to determine the diversity and distribution of the species.

4.4.1.4. Birds

The general objective of the bird research area is to study the temporal evolution and the specific composition of the nesting riparian birds of the river courses as bioindicators of quality status. This is a pioneering initiative as birds are not covered by the WFD in the assessment of the quality status of water bodies.

In this sense, the project aims to establish a sampling protocol in order to specify the bioindicator character during the breeding season of the different bird species by crossing the abundance indices obtained with other factors. These factors include land cover and land use, variables from the sampling of other research areas comprised in the project, or parameters related to the ecology of each species, such as its distribution, abundance or its degree of threat.

The method used for the evaluation of birds is based on the establishment of linear transects with bands, which allow us to calculate the detectability (through auditory and/or visual sensory detection) and to estimate the abundance, taking into account the value of detectability, by applying the distance sampling method (Voříšek et al. 2008).

When evaluating trends over the years, the oscillations of some key or bioindicator species have been studied, especially aquatic birds which are sensitive to environmental alterations, for example the bluefin (*Alcedo atthis*) or the blackbird water (*Cinclus cinclus*), among others.

4.4.1.5. Mammals

In the last few years, this research area has focused on the development of an otter monitoring program with the aim of surveying the return of this species and evaluating its presence and population structure in the study area. The otter is a reliable bioindicator of the recovery of the socio-ecological quality of fluvial systems. The study and population analysis of this mustelid has implications on the management and conservation of riparian ecosystems, as well as playing a key role for social awareness and citizen science initiatives.

The European otter (*Lutra lutra*) has a wide natural geographical distribution, which is a key piece in the ecological balance of river ecosystems. From the second half of the 20th century, the otter suffered a sharp decline in much of Europe. By 1990, it had disappeared from almost all of Catalonia, except in some areas of the Pyrenees. This decline was in response to anthropogenic alteration of river ecosystems, and especially to industrial pollution and its effects on trophic availability (Ruiz-Olmo 2011). From the mid-1990s, the situation improved with the implementation of water treatment measures, and the species has recovered by returning to much of the basins where it had disappeared, including the Besòs and Tordera basins and in the area of the Montseny Biosphere Reserve.

Faced with this new scenario-recent, dynamic, difficult to imagine a few years ago, it is highly relevant to carry out a study that combines the monitoring and analysis of the current situation of this species.

The objectives of the otter monitoring are to analyze and document different aspects of the ecology of the species that are decisive for its establishment and breeding in different sections of recent recolonization. This includes the analysis of the suitability of the habitat, the study of the diet and the determination and evaluation of its distribution.

4.4.1.6. Amphibians and aquatic reptiles

Amphibians are interesting as bioindicators due to their complex life cycle requirements: the presence of aquatic habitats for reproduction and larvae development, as well as a proper terrestrial habitat for their adult life. Moreover, this group is particularly sensitive to global change and is a good indicator of environmental changes (Petitot et al. 2014; Boada et al. 2008).

The main goal of this research area is to evaluate the richness and abundance of the different amphibian and aquatic reptile communities in river courses and adjacent ponds. This evaluation provides information about ecological status, environmental disturbances in fluvial systems and surrounding water bodies, and trends in populations that have been threatened in the last years due to emerging diseases in amphibians. For this reason, biosecurity procedures are adopted in the monitoring protocols.

The applied methodology is based on auditory censuses (night-time call count and visual encounter) to detect and identify species by male calls during spring, during the reproductive period. A listening point is set every 100 meters along a 500-meter transect in the river course, where species identification and number of individuals assessments are conducted during a three-minute period. Moreover, adjacent ponds with high diversity are also evaluated (HDP). In addition to this method, a visual sampling of aquatic reptiles is also performed to detect the grass snake (*Natrix natrix*), the viperine water snake (*Natrix maura*), Mediterranean pond turtle (*Mauremys leprosa*), European pond turtle (*Emys orbicularis*) and non-native species such as the common slider (*Trachemys scripta*).

In reference to anuran, nine species are considered: the Iberian green frog (*Phelophylax perezzi/Phelophyax kl. grafi*), which is the most abundant; the Mediterranean treefrog (*Hyla meridionalis*); the European frog (*Rana temporaria*); the common midwife toad (*Alytes obstetricans*); the parsley frog (*Pelodytes punctatus*); the common toad (*Bufo spinosus*); the painted frog (*Discoglossus pictus*), a non-native species whose population range is expanding; the natterjack toad (*Epidalea calamita*); and the Iberian spadefoot toad (*Pelobates cultripes*). The species of urodelans in the study area are the common fire salamander (*Salamandra salamandra*), the marbled newt (*Triturus marmoratus*) and the palmate newt (*Lissotriton helveticus*).

It is also worth mentioning the populations of Montseny newts (*Calotriton arnoldi*), found in specific areas in the headwaters of the Montseny Biosphere Reserve. This endemic species is considered to be critically endangered (CR) by the IUCN, becoming one of the most endangered amphibians in Europe with specific conservation programs.

4.4.1.7. *Riparian vegetation*

Riparian vegetation, developing in humid and productive soils along the course of rivers and streams, is an essential element of study in assessing the ecological status of river systems.

The role of riparian forests in the dynamics of riparian ecosystems can be defined from a multifunctional point of view, as they have hydrological, ecosystem, landscape and economic implications. The WFD establishes, in Annex V, the quality indicators for the classification of the ecological status, one of which is characterized by the composition and abundance of the aquatic flora.

This research area carries out the assessment of the richness of plant species, their abundance and the hydromorphological structure of the riparian area. Medium- and long-term monitoring of these parameters contributes to the detection of trends in the quality status of riparian ecosystems.

Vegetation inventories are carried out by applying the phytosociological or sigmatist method of Braun-Blanquet (1979), assessing the diversity and abundance of the flora present in three strips of the riparian area: river channel, bank in the ordinary floods area and bank in the large floods area. Non-native species are also considered in the inventories, especially invasive species, as biological invasions are one of the most important components of global change and one of the major threats to conservation for native species according to the International Union for Nature Conservation (IUCN). River systems favor the dispersal of species, including exotic flora, capable of establishing in altered habitats.

4.4.2. *Hydromorphological monitoring*

Hydromorphological monitoring considers the hydrological regime (flow volume, piezometric levels, maintenance flow rates, alteration of the hydrological regime), river continuity and morphological conditions based on the assessment of the riparian vegetation (QBR index).

Observatori Rivus proposes the use of percentiles in relation to the maintenance flow for the assessment of the quality of the annual hydrological regime, using daily averages data.

In relation to the analysis of riparian vegetation as factor for evaluating the riparian area from a hydromorphological point of view, the QBR index is applied as a metric. The QBR (Riparian Forest Quality) index assesses the quality of the riparian forest and the degree of alteration of the riparian zone in four independent blocks: degree of riparian cover, vegetation structure, quality of the cover and naturalness of the river channel (ACA 2006).

4.4.3. Physicochemical monitoring

The organic load or total organic carbon (TOC), salinity (concentration of chlorides and conductivity) and eutrophication (concentration of ammonium, nitrates and nitrites) are evaluated. There is also a chemical water quality assessment based on the analysis of factors such as temperature, pH, oxygen concentration, alkalinity, sulfates, chlorides, calcium, magnesium, sodium, potassium, iron and manganese.

Figure 4.4. *Images of some research areas sampling and methods. From left to right and from top to bottom: sampling of macroinvertebrates and identification of taxa in the laboratory; research of the amphibians and aquatic reptiles; applying biosecurity protocol for material disinfection; electric fishing demonstration; phototrapping technique for otter monitoring*

4.5. Environmental education, communication and training program

Under the premise that rivers are an excellent educative resource, the main objective of PROECA is to transfer scientific knowledge obtained from the different research areas to different spheres of the society by implementing an environmental education and communication strategy. The information flow is considered to be bidirectional, as local knowledge obtained by interviews or participative processes to key actors and local experts is also taken into account and integrated into the applied research described in the previous sections.

Thus, the environmental education and communication strategy aims to promote the dissemination of the results of the project using different channels in accordance with the targeted audience, to generate social interest in relation to fluvial systems, to encourage participation and to promote the discovery of cultural and natural fluvial heritage.

Three scopes are addressed to three main spheres:

– Social: the values of the natural heritage as well as their status of conservation, management practices and the historical and social uses of water are promoted. In this direction, one of the goals of the program is the development of pedagogical materials aimed at different levels of education (primary and secondary) linked to the research areas. These materials, presented during guided fieldtrips to rivers and streams near to the educative center, are designed with the main goal of translating and disseminating protocols of monitoring methodologies and the results obtained to the teaching staff and students.

In terms of general public, communication materials are also produced, such as traveling exhibitions, leaflets or digital contents and audio-visual materials. Moreover, participative sessions are also promoted. The presence in media and social networks is also part of the communication strategy of the project.

– Scientific: the project aims to exchange and disseminate methodologies for the establishment of a monitoring model for river quality status indicators assessment that could be extrapolated to other Mediterranean basins.

In the academic scope, the program includes the training of students and researchers who are undertaking internships, postgraduate studies, master's programs or doctorates from a wide range of different disciplines.

– Political and governmental: the information obtained is synthesized into annual reports which are disseminated to enhance territorial planning and integrated management of water resources in the decision-making processes.

The educational processes that are contemplated in the different scopes according to the target audience are mainly developed in *three areas of action*: formal education, nonformal education and informal environmental education. Thus, based on the different fields of action and the target audience, PROECA develops an environmental education strategy that aims to provide elements for a reconnection of society with its closest fluvial environment (Figure 4.5).

4.5.1. *Formal education*

Formal education is developed in a "school" and academic context. The main objective of formal education corresponds to promoting the learning of various areas of knowledge from the use of didactic and pedagogical materials adapted for each educational level.

4.5.2. *Nonformal education*

Nonformal education contemplates the transmission of knowledge, skills and environmental values outside of the institutional educational system. It entails the adoption of positive attitudes towards the natural and social environment, which are translated into actions of care and respect for biological and cultural diversity and that promote intra- and intergenerational solidarity. It is recognized that environmental education is not neutral, but rather ideological, since it is based on values for social transformation (Declaration of the Land of the Peoples (Foro Río 92 in Castro and Balzaretti 2000)). Nonformal education is carried out outside the curricular contents through organized and systematic activities focused on solving specific learning needs and directed to groups or subgroups of the population.

4.5.3. *Informal environmental education*

Informal environmental education considers actions aimed at training, informing and reflecting on environmental issues through the media such as publications, radio, television or cinema (González 1995 in Castillo 1999).

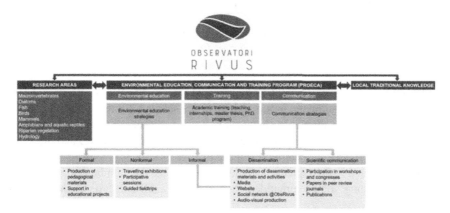

Figure 4.5. *Scheme of the PROECA strategy (source: Maneja 2010). For a color version of this figure, see www.iste.co.uk/romagny/biosphere2.zip*

4.6. A 15-year period implementing PROECA in the Montseny Biosphere Reserve

A data compilation of PROECA activity from 2004 to 2019 shows that up to 101 activities have been conducted in 10 municipalities in the Montseny Biosphere Reserve (31.25% of the total) with 4,786 participants, including scholars and citizens. Overall, there is a clear relationship between the amount of activities and engagement (Figure 4.6; Table 4.1).

Municipalities	Participants	Activities
Sant Celoni	2,113 (44.15%)	57 (56.44%)
Arbúcies	860 (17.97%)	17 (16.83%)
Santa Maria de Palautordera	733 (15.32%)	15 (14.85%)
Breda	430 (8.98%)	3 (2.97%)
Montseny	360 (7.52%)	2 (1.98%)
Riells i Viabrea	154 (3.22%)	1 (0.99%)
Gualba	60 (1.25%)	1 (0.99%)
Sant Esteve de Palautordera	46 (0.96%)	3 (2.97%)
la Garriga	25 (0.52%)	1 (0.99%)
Figaró-Montmany	5 (0.10%)	1 (0.99%)
TOTAL	4,786	101

Table 4.1. *Total number and percentage of participants at the PROECA activities from 2004 to 2019*

Out of the approximately 4,786 participants in the 101 PROECA activities, 1,690 are part of the public audience group and 3,096 are scholars. In that sense, there is a clear difference in the attendance between the two audiences, although the number of activities destined for both audiences is significantly different (62 activities for public audience and 39 for scholars) (Figure 4.7). It must be taken into account that the total number of participants is underestimated, as the number of visitors of exhibitions, among other unguided activities, is difficult to count and is approximate.

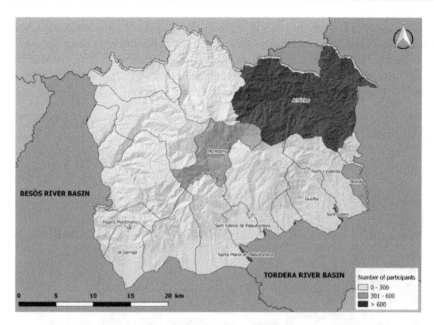

Figure 4.6. *Map of participation at PROECA activities for each of the municipalities of the Montseny Biosphere Reserve in the period 2004–2019. For a color version of this figure, see www.iste.co.uk/romagny/biosphere2.zip*

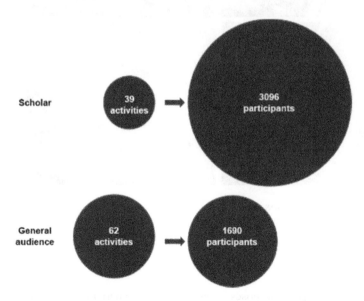

Figure 4.7. *Total number and percentage of participants at the PROECA activities from 2004 to 2019*

Data reveal that over the years, PROECA activities have attracted more scholars than the general audience. Future schedules in the program need to be focused on fostering a better participation of citizens in order to be able to fulfill the objectives of the PROECA and strengthen the relationship between the inhabitants of the Montseny Biosphere Reserve with their own fluvial systems.

Figure 4.8. *Images of environmental education activities with scholars, guided fieldtrips and conferences addressed to general audience*

4.7. Conclusion

Observatori Rivus proposes a cross-cutting experience that encompasses a wide range of activities related to the concept of biosphere reserve as a "learning laboratory for sustainable development". From scientific research for the assessment of the environmental quality status of the Montseny Biosphere Reserve basins, the program aims to transfer knowledge to society through the implementation of an environmental education strategy.

Transferability has been recently implemented from Tordera to Besòs river basin, and it could be replicable also in other Mediterranean territories. In a context of an increasingly informed and empowered society for contribution at the level of research and management, the Observatori Rivus and PROECA experiences demonstrate that biosphere reserves are also suitable scenarios to encourage participation processes and the promotion of citizen science.

As biosphere reserves are considered as science for sustainability support sites, in other words, special areas for testing interdisciplinary approaches to understanding and managing changes and interactions between social and ecological systems, this proposal could become of interest to replicate mid- and long-term monitoring research linked to environmental education and knowledge transfer in relation to fluvial systems.

4.8. Acknowledgements

The authors would like to express their thanks for the support and collaboration that Observatori Rivus has received from local and regional organizations, governmental bodies and private foundations since it was launched in 1996, especially to Fundació Rivus, Consorci Besòs Tordera, Fundació Barcelona Zoo, Diputació de Barcelona, Catalan Water Agency, Consell Comarcal de la Selva and the councils of Sant Celoni, Hostalric, Arbúcies, Santa Maria de Palautordera and Malgrat de Mar.

The authors would also like to thank Antonio Bontempi (Department of Geography, Autonomous University of Barcelona) for his help in reviewing the manuscript, and especially all current researchers involved in the different research areas of the project, as well as all those who have contributed to it throughout this 25-year period.

4.9. References

Agència Catalana de l'Aigua (2006a). BIORI: Protocol d'avaluació de la qualitat biològica dels rius. Report, Departament de Medi Ambient i Habitatge de la Generalitat de Catalunya, Barcelona.

Agència Catalana de l'Aigua (2006b). HIDRI: Protocol d'avaluació de la qualitat hidromor-fològica dels rius. Report, Departament de Medi Ambient i Habitatge de la Generalitat de Catalunya, Barcelona.

Blondel, J., Aronson, J., Bodiou, J.Y., Boeuf, G. (2010). *The Mediterranean Region. Biological Diversity in Space and Time.* Oxford University Press, Oxford.

Boada, M., Mayo, S., Maneja, R. (eds) (2008). *Els sistemes socioecològics de la conca de la Tordera.* Institució Catalana d'Història Natural, Barcelona.

Braun-Blanquet, J. (1979). *Fitosociología. Bases para el estudio de las comunidades vegetales.* Blume, Madrid.

Consell Assessor per al Desenvolupament Sostenible de la Generalitat de Catalunya, Institut d'Estudis Catalans (2017). Tercer informe sobre el canvi climàtic a Catalunya. Report, Generalitat de Catalunya, Barcelona.

Corraliza, J.A., Collado, S., Bethelmy, L. (2013). Spanish version of the new ecological paradigm scale for children. *Spanish Journal of Psychology*, 16(27), 1–8. doi: 10.1017/sjp.2013.46.

Cudennec, C., Leduc, C., Koutsoyiannis, D. (2007). Dryland hydrologyin Mediterranean regions. *Hydrological Sciences Journal*, 52, 1077–1087. doi: doi.org/10.1623/hysj.52.6.1077.

European Community (2000). Directive 2000/60/EC of October 23 2000 of the European Parliament and of the Council establishing a framework for community action in the field of water policy. *Official Journal of the European Union*, 327, 1–72.

García-Berthou, E., Bae, M.J., Benejam, L., Alcaraz, C., Casals, F., de Sostoa, A., Solà, C., Munné, A. (2016). Fish-based indices in Catalan rivers: Intercalibration and comparison of approaches. In *Experiences from Surface Water Quality Monitoring: The EU Water Framework Directive Implementation in the Catalan River Basin District*, Munné, A., Ginebreda, A., Prat, N. (eds). Springer, Berlin.

Gomà, J. (2005). Metodologia per a l'estudi de les diatomees a la conca de la Tordera. Report, L'Observatori, Estació de seguiment de la conca de la Tordera.

Gudmundsson, L. and Seneviratne, S.I. (2016). Anthropogenic climate change affects meteorological drought risk in Europe. *Environmental Research Letters*, 11(4), 45. doi: 10.1088/1748-9326/11/4/044005.

Hermoso, V. and Clavero, M. (2011). Threatening processes and conservation management of endemic freshwater fish in the Mediterranean basin: A review. *Marine and Freshwater Research*, 62(3), 244–254. doi: 10.1071/MF09300.

James, L.A. and Marcus, W.A. (2006). The human role in changing fluvial systems: Retrospect, inventory and prospect. *Geomorphology*, 79, 152–171. doi: 10.1016/j.geomorph.2006.06.017.

Lorenz, A.W., Stoll, S., Sundermann, A., Haase, P. (2013). Do adult and YOY fish benefit from river restoration measures? *Ecological Engineering*, 61, 174–181. doi: 10.1016/j.ecoleng.2013.09.027.

Maneja, R. (2010). La percepción del medio ambiente en grupos infantiles y adolescentes. Comparativa entre la Huacana (Michoacán, México) y la cuenca del río Tordera (NE, Cataluña). PhD Thesis, Autonomous University of Barcelona [Online]. Available at: https://www.educacion.gob.es/teseo/imprimirFicheroTesis.do?idFichero=wHG%2FqxXPdqc%3D.

Mediterranean Experts on Climate and Environmental Change (2019). Risk associated to climate and environmental changes in the Mediterranean region. A preliminary assessment by the MedECC network. Science-policy interface-2019 [Online]. Available at: https://www.medecc.org/.

Otero, I., Boada, M., Badia, A., Pla, E., Vayreda, J., Sabaté, S., Gracia, C.A., Peñuelas, J. (2011). Loss of water availability and stream biodiversity under land abandonment and climate change in a Mediterranean catchment (Olzinelles, NE Spain). *Land Use Policy*, 28(1), 207–218. doi: 10.1016/j.landusepol.2010.06.002.

Petitot, M., Manceau, N., Geniez, P., Besnard, A. (2014). Optimizing occupancy surveys by maximizing detection probability: Application to amphibian monitoring in the Mediterranean region. *Ecology and Evolution*, 4(18), 3538–3549. doi: 10.1002/ece3.1207.

Rosenberg, D.M. and Resh, V.H. (eds) (1993). Introduction to freshwater biomonitoring and benthic macroinvertebrates. In *Freshwater Biomonitoring and Benthic Macroinvertebrates*. Chapman & Hall, New York.

Ruiz-Olmo, J. (2011). Factors affecting otter (Lutra lutra) abundance and breeding success in freshwater habitats of the northeastern Iberian Peninsula. *European Journal of Wild-Life Research*, 57(4), 827–842.

Sánchez-Hernández, J. and Amundsen, P.A. (2015). Trophic ecology of brown trout (*Salmo truta* L.) in subarctic lakes. *Ecology of Freshwater Fish*, 24, 148–161. doi: 10.1111/eff.12139.

Sànchez-Mateo, S. and Boada, M. (2011). Anàlisi socioecològica del Vall de Santa Fe (massís del Montseny): La transformació del paisatge a través de la història ambiental [Online]. Available at: https://ddd.uab.cat/record/98755?ln=ca.

Simon, T.P. (ed.) (1999). Introduction: Biological integrity and use of ecological health concepts for application to water resource characterization. In *Assessing the Sustainability and Biological Integrity of Water Resources using Fish Communities*. CRC Press, Boca Raton.

Vega, M., Pardo, R., Barrado, E., Deban, L. (1998). Assessment of seasonal and polluting effects on the quality of river water by exploratory data analysis. *Water Research*, 32(12), 3581–3592. doi: 10.1016/S0043-1354(98)00138-9.

Voříšek, P., Klvaňová, A., Wotton, S., Gregory, R.D. (eds) (2008). A best practice guide for wild monitoring schemes. Report, CSO/RSPB.

Voza, D., Vuković, M., Takić, L.J., Nikolić, D.J., Mladenović-Ranisavljević, I. (2015). Application of multivariate statistical techniques in the water quality assessment of Danube River, Serbia. *Archives of Environmental Protection*, 41(4), 96–103. doi: 10.1515/aep-2015-0044.

5

Do We Need to Choose Between Biodiversity, Industry and Tourism? A Metabolic Approach to Manage the Mediterranean Biosphere Reserve of Menorca

5.1. Introduction

Both the 1971 Man and the Biosphere (MAB) Program and the 2015 Sustainable Development Goals are examples of international responses to make a decent human life and development compatible with the biosphere and the use of resources. During these 50 years, sustainability issues have increased their influence in the international agenda. Five "Earth Summits" – Stockholm 1972, Nairobi 1982, Rio de Janeiro 1992 and 2012, and Johannesburg 2002 – the Millennium Development Goals (2000) and other global and regional initiatives prove this interest and increasing relevance of sustainability. However, is the world more sustainable today than it was 50 years ago? While these initiatives and goals provide a valuable contribution and guidelines for several regions to include sustainability in their agendas, some issues arise: what are the limits of regional decision-making? How do we account for environmental and social impacts happening out of a specific region? Can a region be sustainable without considering the impacts it generates somewhere else? Are target-oriented initiatives the best governance tool to approach sustainability issues? To what extent can different voices be included in the debate, deliberation and decision-making when there are contradicting goals, interest and outcomes?

Chapter written by Alejandro MARCOS-VALLS.

These questions entail the uncertainty of dealing with complex issues when it is nearly impossible to control everything and still decisions need to be taken. The emergence of this complexity in society requires the development of a post-normal science perspective beyond the reductionist approach assumed by traditional science (Funtowicz and Ravetz 1994). Post-normal science responds to high levels of uncertainty or decision stakes (Funtowicz and Ravetz 1993) and highlights the need for transparent, flexible and adaptive approaches to not necessarily provide clear answers, but contextualize information and reference it to different scales and dimensions to make conscious decisions while being open to change. We argue that societal metabolism approaches, and specifically MuSIASEM (Multi-Scale Integrated Analysis for Societal and Ecosystem Metabolism), provide flexible and adaptive tools to contribute to coordinate policies that improve the quality of life of the citizens in a region while preserving biodiversity, other natural resources and the social fabric inside and outside the specific regions we are managing.

This chapter aims to explain how the use of a metabolic approach applied to socio-ecological systems makes it possible to perform an integrated analysis for decision-making that is flexible and transparent. After introducing the concept of societal metabolism and the basic rationale behind the MuSIASEM approach, we will focus on the case of the Mediterranean island of Menorca (Spain), which has been a biosphere reserve since 1993. We will contextualize the situation of the Menorca Biosphere Reserve and will briefly introduce its Action Plan for 2025, the adaptation of the Sustainable Development Goals and Agenda 2030 to the island. Then, we will illustrate the application of MuSIASEM to Menorca and discuss how metabolic indicators can open spaces for deliberation among stakeholders about potentially contradicting goals and interest and aid the decision-making process.

5.2. Societal metabolism

The concept of metabolism is usually associated with the activity of a living organism and the processes taking place in order to maintain life. However, the basic rationale of systems that require energy and matter to achieve its internal order and (re)produce themselves has also been applied to socio-ecological systems (Ostwald 1907, 1911; Lotka 1922, 1956; Soddy 1926; Zipf 1941; White 1943; Cottrell 1955). This application has generated an emerging field of study, "societal" (or social) metabolism, that characterizes socio-ecological systems regarding the use of energy and matter and the production of wastes and emissions (Martinez-Alier 1987; Fischer-Kowalski 1998; Fischer-Kowalski and Hüttler 1998; Daniels and Moore 2002; Giampietro Mayumi and Ramos-Martin 2008; Martinez-Alier et al. 2010; Giampietro et al. 2014).

Despite the specific characteristics of each socio-ecological system – meteorological, geographic, geological and socio-economic, among others – globalization is producing common pressures and changes that are modifying their functional and structural organization. By societal metabolism, we refer to the resource and energy inflows, and waste outflows that are necessary to maintain and reproduce the functions and structures of the system. The control and monitoring of the changes experienced by the system, and the better understanding of changes in the structure of their metabolism, require new policies and new analytical tools. To deal with ever-increasing complexity, we need analytical tools and institutions that can handle such complexity.

Several researchers have presented metabolic approaches to understand and explain what happens in diverse socio-ecological systems and to assist policy-making (Fischer-Kowalski 1998; Giampietro and Mayumi 2000; González de Molina and Toledo 2014; Fraňková et al. 2017). From urban metabolism (Wolman 1965; Swyngedouw 2006; Broto et al. 2012; Pérez-Sánchez et al. 2019) to the metabolism of islands (Singh et al. 2001; Martínez-Iglesias et al. 2014; Ginard and Murray 2015; Bogadóttir 2020), including regions and states (Carpintero 2005; Ginard and Murray 2015; Velasco-Fernández et al. 2015; Velasco-Fernández 2017), there are examples of the application of the concept. In this chapter, we will focus on the application of an accounting system called Multi-Scale Integrated Analysis of Societal and Ecosystem Metabolism (MuSIASEM).

5.3. MuSIASEM: integrating information from multiple scales to improve participation and stakeholder engagement

The MuSIASEM approach is an accounting methodology that makes it possible to establish a bridge across different indicators defined across different levels and dimensions (Giampietro 2014). The use of indicators at different levels maintains multi-dimensionality and extra complexity; if we use one indicator and one level, we simplify. More specifically, MuSIASEM provides a set of tools to identify the factors determining the quantity and the quality of input and output flows of energy and materials by establishing a relation between the analyzed flows and the functional and structural elements of the system under study, based on Georgescu-Roegen's (1975) flow-fund model. Flows refer to elements that appear or leave the system over the duration of the representation – for example fossil energy – and funds refer to elements that remain over the duration of the representation and transform input flows into output flows (Giampietro et al. 2008).

One of the key aspects of MuSIASEM is that it is based on the hierarchical idea of a holon (Koestler 1967, 1979). A holon as a system that can be understood as a whole, as a part of bigger systems, and containing other (sub)systems, that in MuSIASEM are defined as hierarchical levels of analysis. The existence of different levels of analysis enables us to understand and visualize integrated information, bearing in mind the relation to the reference system. The usefulness of having a multi-scale perspective is that it allows for contextualizing information and eases stakeholders from different levels of analysis to participate in adding data and understanding the relation of these data with bigger or smaller systems. The possibility of defining the size of the system and the scale of analysis is an important feature of MuSIASEM, and it provides a flexible and adaptive approach to sustainability issues.

Even though it is not necessary to carry out all of them to apply MuSIASEM, the approach includes three stages: 1) the definition of the system, 2) the diagnosis and characterization of the societal metabolism and 3) the anticipation and simulation of different scenarios (Marcos-Valls et al. 2020). The definition stage implies deciding *what is the system* that we want to analyze – constituent components, hierarchical organization of the relations over functional and structural elements across the different levels, and the metabolic characteristics (flow/fund ratios) – and *what kind of data* are available and how to obtain it. The diagnosis stage allows us to understand who uses certain types of resources for what activities in order to maintain and reproduce the system. This stage includes an overview of the degree of dependency or self-sufficiency in relation to other systems, and the potential environmental pressures and impacts connected to the considered flows (electricity, fuels, blue water, grains, etc.). Finally, the anticipation stage allows us to develop scenarios to discuss and deliberate about the implications of alternative scenarios in terms of the metabolic performance, dependency of external resources or environmental impacts linked to maintaining and reproducing the system as it is or modifying it. The three stages reflect three critical questions introduced to operationalize how to deal with sustainability: sustainability of what? (Stage 1. Defining the system); sustainability for whom? And why? (Stage 2. Diagnosis) (O'Connor 2006; Frame and O'Connor 2011) and finally sustainability for how long? (Stage 3. Anticipation).

The MuSIASEM approach includes the complexity of dealing with feasibility and viability, in biophysical terms, and desirability, in social and cultural terms, of alternative scenarios. Feasibility, regarding the availability of resources depending on processes outside human control. Viability, when it comes to the possibility to implement alternatives of use and transformation of resources (use of different technologies, carriers, etc.) in processes under human control. And desirability of alternative policies considering the consequences at different scales on the quality of life, expectations and normative values of the society. Hence, the use of integrated approaches, such as MuSIASEM, can enable participation and stakeholders' engagement to contrast scenarios and analyze how viable and feasible desirable/sustainable alternatives are, as suggested by stakeholders with different perceptions and interests.

Before we present the case of Menorca and the application of MuSIASEM to the island, the following sections will contextualize biosphere reserves, the Sustainable Development Goals (SDGs), and its role on the Mediterranean region and particularly on Menorca.

Figure 5.1. *Biosphere reserves and the Sustainable Development Goals-Agenda 2030. Source: Sònia Sànchez-Mateo*

Under the MAB Program, biosphere reserves are an instrument to experiment how to reconcile the conservation of biodiversity with sustainable practices. A total of 701 sites in 124 countries (August 2020) constitute a dynamic worldwide network that aims to build "thriving societies in harmony with the biosphere" (UNESCO 2020) balancing policies with scientific and academic knowledge and citizen participation.

Biosphere reserves appeared as a response to a series of environmental problems, a consequence of unsustainable models of development. Since 1971, when the MAB Program was launched, the World Network of Biosphere Reserves have increased in numbers and also in experiences and in co-created knowledge to integrate human activities with nature. Biosphere reserves are supported by international, regional, sub-regional and ecosystem-specific networks to facilitate the learning among related reserves and the exchange of experiences and initiatives. The unequal situation of biosphere reserves reinforces the need for sharing mechanisms that allow biosphere reserves with fewer resources to easily adapt successful experiences from others within the network. Therefore, other centers and institutions exist to stimulate exchanges and disseminate experiences among biosphere reserves with common characteristics, such as the International Center on Mediterranean Biosphere Reserves (UNESCO 2014).

Beyond the exchange of collective knowledge to implement more sustainable policies and management practices among the different biosphere reserves-related networks, these experiences are crucial to develop sustainability strategies for the rest of the planet to implement, such as the Sustainable Development Goals (SDGs) or the Millennium Development Goals before.

In 2015, the UN Member States adopted the Sustainable Development Goals as part of the 2030 Agenda for Sustainable Development (United Nations 2015) to promote sustainable practices and policies integrating resource uses, environmental and biodiversity protection, and social, economic and well-being promotion. The 17 goals and 169 targets entail an enormous international effort to achieve an agreement to implement these goals aiming "peace and prosperity for people and the planet". However, there are already some critical voices questioning the consistency of the SDGs, and its targets and indicators, as a way of depoliticizing sustainability issues using strong narratives (Spaiser et al. 2017; Menton et al. 2020; Velasco-Fernández 2020) that neutralize alternative narratives and thus potential solutions.

The complexity around sustainability and governance requires addressing the issue of how to prioritize indicators and targets referring to nonequivalent policy concerns, which is a normative decision and not a scientific one (Giampietro 2019), and this depends on who controls the process. The monitoring of indicators and targets independently can give us the wrong idea of success, creating new sustainability problems that we have not even considered. The interactions among targets and indicators are complex and a systemic approach is needed to identify and address potential side effects and trade-offs (Weitz et al. 2018). The use of indicators for governance allows us to justify decisions because there are some previous agreements or acceptance of the process: *"science speaks the truth to follow"*. However, this is a problematic simplification in complex systems where there is a plurality of nonequivalent perspectives and legitimate values that cannot be measured and compared (Kovacic and Giampietro 2015). The claim for integrated analysis and reflexivity is connected to the fact that there is not a single truth but several truths that need to be acknowledged. Different stakeholders will carry different perspectives and representations about the issues at stake and therefore integrated and reflexive ways of framing, discussing and deciding about the issues are crucial to create a trustworthy and transparent mechanism to include different voices and concerns. In this context, if complexity is ignored, or hidden, it is because one dimension, goal or target (typically the economic one) is being privileged. This is not a scientific choice, it is a normative choice, and normative choices should happen in the democratic sphere and not indirectly by hiring technicians that privilege one domain over the others.

The SDGs provide a global framework for sustainable development, but the dominant structure around indicators and targets require a political discussion that needs to be context-specific (Giampietro 2020). Biosphere reserves present the appropriate scale and provide a case-by-case approach to decide the relevant targets and indicators in order to prioritize where direct stakeholders can participate and contribute. However, there is a need to adapt decision-making processes to complexity and establish integrated methodologies and ways of framing, discussing and deciding about the issues at stake at different levels. To promote effective policies for sustainability, the impacts of these policies must be considered inside and outside the system to manage them, and to include and empower stakeholders to participate and trust the quality of the process even with high levels of uncertainty.

5.4. The case of Menorca: a Mediterranean Biosphere Reserve with an action plan to implement the sustainable development goals

The Mediterranean region concentrates an important number of biosphere reserves and involves the participation of several countries. For decades, biosphere reserves across the Mediterranean have been designing, testing and establishing different management practices to balance human development with environmental protection. These experiences offer many lessons to learn from and new questions to deal with.

The Balearic island of Menorca (Spain) has been a biosphere reserve since 1993 (UNESCO 1993), extended in 2004 and 2019. During this period, Menorca has performed different functions related to the biosphere reserve through CIME (Menorca Island Council), IME-OBSAM (Menorcan Studies Institute – Socio-Environmental Observatory) and the participation of civil society and other entities. On the one hand, it has presented several initiatives and plans for its sustainable development (IDS and IME 1996; Cardona Pons et al. 2018; CIME and Ezquiaga 2019; Consell Insular de Menorca 2019), and it is also continuously monitoring and reporting social, economic and environmental indicators (Canals Bassedas et al. 2013; IME and CIME 2019; OBSAM 2020). Moreover, Menorca has developed an action plan for 2025 based on the Agenda 2030 in order to reach the targets and indicators defined by the SDGs (Consell Insular de Menorca 2019).

Figure 5.2. *Dune system and associated vegetation.*
Source: Sònia Sànchez-Mateo

This section focuses on Menorca, using it to illustrate how the concept of societal and ecosystem metabolism, and specifically the Multi-Scale Integrated Analysis of Societal and Ecosystem Metabolism (MuSIASEM) approach, can integrate complex information to complement the implementation of the SDGs or other initiatives for sustainability beyond targets and goals, and the island system. How do managers in Menorca deal with biodiversity? How does the Action Plan 2025 consider biodiversity to meet the SDGs connected to the Agenda 2030? To what extent are there contradicting indicators and targets? And, how are they managed and prioritized? How is biodiversity negotiated with other activities and needs? To what extent can a metabolic perspective, such as MuSIASEM, provide scientific information that includes different stakeholders to participate in the discussion? In the following sections, we will try to discuss these and other questions.

First, we will explore Menorca's Action Plan 2025 to implement the SDGs with special attention to references to biodiversity. Then, we will show some results of the application of MuSIASEM to Menorca, focusing in the metabolic performance of the economic sectors: Agriculture; Industry, Construction and Manufacturing; and Services and Government. We want to understand i) to what extent the plan is dealing with contradictory goals related to biodiversity and ii) how the use of a metabolic approach can contribute to a more inclusive, transparent and adaptive way to deal with sustainability issues in the Mediterranean and island context.

5.5. Menorca 2025. An Action Plan for the Menorca Biosphere Reserve

In 2019, the Action Plan for the Menorca Biosphere Reserve was approved (Consell Insular de Menorca 2019). The document develops a complete strategy to promote a sustainable development based on the conceptual framework of the SDGs and the requirements of the MAB strategy 2015–2025 (UNESCO 2017) the Lima Action Plan 2016–2025 (UNESCO 2016), and the Ordesa-Viñamala Action Plan 2017–2025 (MaB España and OAPN 2017) under the guidelines of the Seville strategy (UNESCO 1996). In its Summary Document (CIME and GOIB 2019), the Action Plan recognizes that the trend between the development of the economy and the conservation of nature was unsustainable (ibid) and therefore new actions are needed to promote sustainability.

The Action Plan presents four lines of action, six main objectives with several operational objectives organized in different Action Programs. We will contrast the two main objectives from the plan connected to different SDGs: main objective 1 and main objective 3 (Figure 5.1). We will focus on the reference to the word "biodiversity" in the operational objectives and actions. If the word does not appear, such as in the case of the main objective 3 or any of its operational objectives, we will discuss the potential impacts of these actions and objectives towards biodiversity.

Main objective 1 responds to the "Programme for the conservation and sustainable use of natural heritage, biodiversity and landscape", which is related to SDGs 3, 6, 11, 14 and 15 (3. Good health and well-being; 6. Clean water and sanitation; 11. Sustainable cities and communities; 14. Life below water; and 15. Life on land). The main objective 3 is included in the "Sustainable economy promotion programme", related to SDGs 1, 2, 8, 9, 10, 12 (1. No poverty; 2. Zero hunger; 8. Decent work and economic growth; 9. Industry, innovation and infrastructure; 10. Reduced inequalities; and 12. Responsible consumption and production).

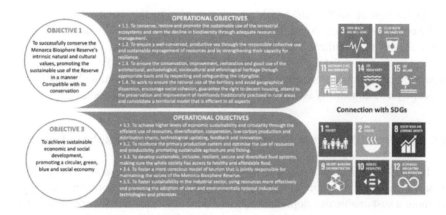

Figure 5.3. *Selected main objectives, operational objectives and connection to the SDGs from the Action Plan for the Menorca Biosphere Reserve (adapted from CIME and GOIB (2019)). For a color version of this figure, see www.iste.co.uk/romagny/ biosphere2.zip*

Rather than an exhaustive analysis of goals and actions, we want to reflect on the need to look beyond the goals and its achievement as solutions, and discuss an integrated approach to deal with the complexity and uncertainty that sustainability embodies.

By using examples from this complete Action Plan based for the Menorca Biosphere Reserve, we would like to open the debate about current global tools to achieve sustainable development and to articulate policies around economic development and environmental protection, such as the SDGs. The complexity of the endeavor implies a risk for paralysis among decision-makers either "by analysis" of too much information (Maxim and van der Sluijs 2007), "by precaution" to avoid uncertain consequences (Cooney 2006) or "democratic paralysis" (Kovacic 2015) and therefore the existence of plans and strategies facilitate actions, even though sometimes the search for specific solutions can cause undesirable outcomes. We would like to offer an integrated perspective where the discussion moves from solution-based approaches to a process-based approach that is inclusive, flexible and adaptive where stakeholders can discuss alternatives to achieve a goal and what are the implications for sustainability of dealing with these alternatives.

The "program for the conservation and sustainable use of natural heritage, biodiversity and landscape" focuses its attention on one of the three pillars of the biosphere reserve declaration: how to make compatible the use of resources with its conservation. On the other hand, the "sustainable economy promotion program" centers the attention on the "develop[ment of] activities that are environmentally and socially responsible, as well as financially viable and profitable" (CIME and GOIB 2019) using circular economy as a main conceptual tool. Is it possible? To what extent can a society do the same, or even more, with less resources? What role does biodiversity management have in this case? The following section will introduce the metabolic performance of the economic sectors in Menorca in order to be able to evaluate potential trade-offs and side effects of promoting an economic activity in one sector over another.

5.6. Metabolic performance of economic sectors in Menorca. Application of the MuSIASEM approach

In the following paragraphs, we will present a part of the diagnosis of the societal metabolism of Menorca mainly based on the metabolic performance of the three economic sectors in the island for the year 2015 (Marcos-Valls et al. 2020).

Figure 5.2 represents the results of metabolic performance of the system. The metabolic performance is an indicator of the quantity of resources and emissions needed to produce a euro of gross value added (GVA), and the resources, emissions and value added attached to a square meter of land use (LU), and to an hour of human activity (HA) in each economic sector. In this case, we focus on the use of energy, considering the thermal equivalent of different energy types (ET) (including electricity, fuels and heat), blue water (WT), and on direct CO_2 emissions. This intensive information is complemented by the extensive values of how the total amount of hours of human activities (THA), hectares of land (TLU) and gross value added (TGVA) are distributed among economic sectors (see Figure 5.2). The size of the sections in the figure is proportional to the higher value per each row and color.

It is important to note that more than the actual numbers, the metabolic performance shows relations and ratios of resources used and emissions produced in order to perform these activities. For the purpose of this chapter, we are only showing one scale: the economic sectors, and land uses and human activities devoted to noneconomic activities, such as household activities or other nonpaid activities, are not represented in the figure. However, the hours of human activity devoted to households in Menorca, including temporary and permanent residents in 2015, are up to 919 Mh when compared to the time devoted to activities in the economic sectors (see THA in Figure 5.2) that represents almost 95% of the total time. A similar distribution of LU shows that households occupy 1.920 ha, approximately 3% of the total, and other land uses that are not directly managed by an economic sector, including natural areas, represent 37.180 ha, more than half of the total land of Menorca.

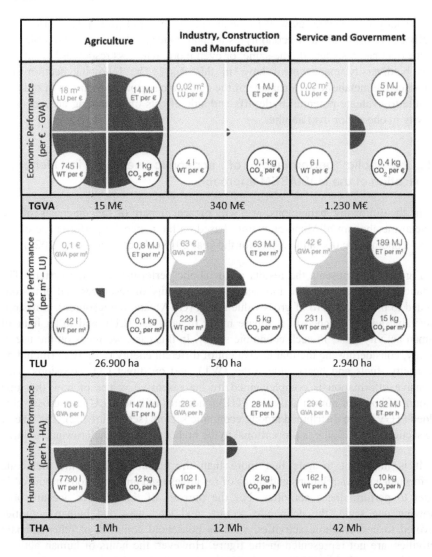

Figure 5.4. *Metabolic performance and extensive variables of the three economic sectors in Menorca 2015 (adapted from Marcos-Valls et al. (2020)). For a color version of this figure, see www.iste.co.uk/romagny/biosphere2.zip*

Even though it is not applied in this chapter, the methodology allows us to upscale the level of analysis to all of the economic activities as "paid work" or downscale each economic activity to see how different subsectors or specific economic activities perform in metabolic terms: for example, comparing agricultural activities with fishing or livestock, or different types of crops or cattle. This feature of the methodology, based on the idea of holon presented above, brings transparency and allows different stakeholders to include new data when it is available or modify/correct information when discussing and deliberating about potential policy alternatives.

Results show an important divide between industry and services sectors, on the one hand, and agricultural activities, on the other hand. While Industry, Construction and Manufacture (ICM) and Services and Government (SG) are able to generate more value added with less resources per each euro generated, we see that the volume of the value added in both activities, especially Services and Governance, that include transportation, implies that in absolute terms both activities use much more resources than Agriculture. Agriculture has a higher dependency of local resources, especially water and land, while industry and services outsource the use of resources, externalizing part of the environmental pressure and impact.

If we look at the Land Use performance – euros generated, resources used or emissions produced per unit of land (m^2) – we see a similar divide. While agricultural activities have little impact per unit of land, the fact that occupies a big share of the land in Menorca amplifies the need of resources, especially water, while it still keeps a low economic value added. Industry and services present a high resource use and emissions per each squared meter and high value added, due to the fact that both activities occupy less land in the island. The amount of energy (ET) needed for squared meter of services and governments is especially high because it includes transportation and a direct use of fuels for this activity.

The analysis of the euros generated, resources and emissions linked to an hour in each economic activity, the Human Activity performance, reveals a similar pattern. Agriculture, even representing a small share of the total hours of human activity, requires a significantly high amount of water per each hour of work, more than 70 times greater than industry and almost 50 times greater than services, while producing a third of value added. The total energy and emissions show also higher relative values, but the difference of hours devoted to industry and especially to services reduces the absolute impact.

5.7. Discussion: do we need to choose between biodiversity, industry and tourism?

At the beginning of the previous section, we introduced a few questions related to how biodiversity is managed in the Menorca Biosphere Reserve in relation to the SDGs and Agenda 2030 and the different issues that may appear in this management. The Action Plan for the Menorca Biosphere Reserve 2025 introduces a few main objectives that the island is following in order to meet the SDGs. The objectives and specific actions aim to be concrete and relatively easy to measure, but they are often too open and defined in such a way that the action can be formally accomplished without evaluating its impact on biodiversity or its contribution to sustainability on a bigger scale.

Considering biodiversity in policy-making implies an awareness of the interactions and trade-offs among different human activities that take place in a society and the environmental impacts it generates. Knowledge about these interactions and impacts is often partial and implies the use of different narratives and epistemologies that not necessarily recognize or understand each other. Therefore, managers and policy-makers are often walking on a fine line that implies promoting actions that they may find desirable for their areas of expertise, interest or the region where they work but that can create undesirable effects at different scales.

The Action Plan for the Menorca Biosphere Reserve shows an ambitious set of actions to protect biodiversity with a program that includes objectives aiming to "conserve, restore and promote a sustainable use of terrestrial ecosystems" (1.1), to "ensure a well-conserved coastline and productive sea" (1.2) and work "to ensure the rational use of the territory" (1.4). At the same time, the action plan includes a program to promote a sustainable economy with objectives aiming at "achieving economic sustainability and circularity" (3.1.), "developing sustainable, inclusive, resilient, secure and diversified food systems" (3.3) or fostering "a more conscious model of tourism" (3.4) or "sustainability in the industrial sector through more effective resources [...] and clean and environmentally rational technologies" (3.5).

As already mentioned, the specific actions to fulfill these objectives are often open – that is, "application of good practices in all activities organized in the Reserve" (A.1.1.9) or "promotion of a form of ecotourism that will help shift the paradigm towards a sustainable tourism model" (A.3.4.3). It is therefore difficult to say whether the accomplishment of the action will protect biodiversity or promote sustainability, although they are designed to achieve this goal. The use of a metabolic perspective suggests a new logic to deal with these issues. The consideration of potential trade-offs among different alternatives will allow us to develop scenarios and to assess the consequences of those alternatives. The development of scenarios based on relational information about the resources needed to develop certain functions and maintain structures is crucial to develop transparent policies for sustainability and to check how certain actions interact among themselves.

The understanding of the metabolic pattern and the performance in the use of resources of the system makes it possible to consider relevant information related to what it means to promote certain human activities over others in terms of time, land and economic values in a system. As the results show, high economic benefit is often linked to an open system in terms of energy sources, and high pressure on water resources, such is the case of industry (ICM). Therefore, promoting industry can increase the pressure over resources and/or the dependence on external resources. The integrated analysis of the use of resources and production of emissions in relation to human activity, land use and the value added provides a more complete vision of the system. In this way, it is possible to study trade-offs among alternatives once they are decided: it is not the same to promote the yachting sector instead of footwear (3.5), and therefore we need to study and compare the metabolic indicators of both industries before making a decision.

As we saw above, agriculture depends on local resources and has a low contribution to the value added directly for its main function, which is food provision. The analysis does not consider other functions provided by agriculture, such as keeping traditional landscape management practices and other traditions. However, it is important to understand how this multi-functionality contributes to the generation of wealth in other sectors, for example, how agricultural land management contributes, through the provision of cultural services to tourism for example, or to increase biodiversity by promoting traditional agricultural land patches in Menorca.

The same applies in the opposite direction. The promotion of sustainable tourism can be seen as a desirable activity since the services sector has a higher economic performance and promotes a sustainable and respectful behavior towards nature and the environment of Menorca. However, it is important to know how far resources are externalized for this tourist to arrive to the island, or to know how many tourists will take a certain route because it is popular this year, the other routes are closed or any other reason, affecting the biodiversity in this area.

The externalization of the demand of resources and the environmental pressures and impacts associated with this externalization implies that the promotion of this kind of economic activities generates a more sustainable development locally, shifting the environmental and social pressure to other places, which should be taken into account when discussing policies and implementing actions for a sustainable development.

As we can see in the results of the study, we can obtain information about how the economic activities perform in a system. We can also define goals and targets that we believe are desirable and beneficial for (a) society but decision-making implies also to prioritize one alternative over another with the participation of different stakeholders and considering information that may come from different sources. Just considering one part of the information can lead to decisions that may be desirable for some and undesirable for others.

The adaptation to the SDGs to the Menorca Biosphere is a strategic exercise to evaluate the situation and adopting actions and goals to aim for a more sustainable society. However, the complexity of sustainability issues and the interconnection between actions and consequences all around the globe make it difficult to achieve sustainability based on these actions and goals. The sustainability of a social–ecological system requires knowledge of the relationships between resources, use and disposal at different stages, scales and places.

The integrated analysis provides transparency about trade-offs and provides a structure that shows how changes can modify the metabolic performance across the system. This in combination with the participation of different stakeholders makes it possible to create a democratic and deliberative process to contrast and share data, alternatives and the implications of implementing certain actions over others.

5.8. Conclusion

Since its declaration as a biosphere reserve in 1993, Menorca has developed several initiatives to promote sustainable development and balance human activity with nature. The island has approved its Action Plan for the Menorca Biosphere Reserve to meet the SDGs and is planning to implement its actions until 2025. The document presents ambitious actions with a calendar and required budget to achieve the defined objectives. Conservation of natural heritage, biodiversity and landscape; efficient resources management; environmental education; and monitoring of the acceptable impacts are the four lines of action included in the plan. The operational objectives and actions are designed to be connected to the SDGs and every program connects with some of the 17 goals.

The difficulties and limitations of considering biodiversity and sustainability issues in a goal-oriented manner in biosphere reserves, such as the presented Action Plan for the Menorca Biosphere Reserve, lie in the fact that achieving a goal allows for different solutions that normally are not integrated or cohesive with solutions for other goals that can interact among themselves. This chapter shows an example of how integrated approaches, such as MuSIASEM, enable us to deal with complexity and contextualize statistical and numerical data in a transparent way. The approach requires us to be open about exploring alternative scenarios and tries to minimize problems related to reductionism and the use of partial scientific data, which can lead to decisions against the general interest if the complexity of the system is not considered (Giampietro and Mayumi 2018).

Menorca is an open socio-ecological system, with a high dependency on external resources. This analysis shows that services (SG) and industry (ICM) contribute directly to the economy in Menorca more than agriculture (AG), far behind. Both services and industry depend on external resources and use more resources per unit of area, while agriculture depends on local resources, such as water. These results can inform policy-making to decide which economic models are to be promoted or how economic sectors need to transform to adapt to changing biophysical and social constraints.

The high dependency on external resources of Menorca shows that most of the environmental pressures and impacts are located outside of the biosphere reserve. Therefore, any strategy or action for sustainability should also balance its impact out of the boundaries of the biosphere reserve. Dealing with biodiversity, directly linked to the territory, makes it an extra degree of complexity since managers and policy-makers need to take concrete and localized actions to preserve biodiversity on-site while, at the same time, they should care about the contribution of these actions to global changes, such as climate change, that could affect biodiversity in the long run.

Biosphere reserves are already at the forefront of implementing policies for sustainable development and maintaining a balance between human development and nature. After years of experience and a lot of shared knowledge about initiatives, biosphere reserves are in a position to explore the way to new strategies and approaches to integrate systemic information to understand what is happening in the reserve and also the implications of these actions for global sustainability.

5.9. References

Bogadóttir, R. (2020). The social metabolism of quiet sustainability in the Faroe Islands. *Sustainability*, 12(2), 735. doi: 10.3390/su12020735.

Broto, V.C., Allen, A., Rapoport, E. (2012). Interdisciplinary perspectives on urban metabolism. *Journal of Industrial Ecology*, 16(6), 851–861. doi: 10.1111/j.1530-9290.2012.00556.x.

Canals Bassedas, A. and Carreras Martí, D. (2013). 20 anys d'una il·lusió. Jornades sobre els 20 anys de la reserva de biosfer de Menorca, Jornades sobre els 20 anys de la reserva de biosfera de Menorca. Report, Institut Menorquí d'Estudis, Agència Menorca Reserva de Biosfera, Maó.

Cardona Pons, J., Camps Orfila, X., Pons Maria, M. (2018). Menorca's first energy transition. Energy system analysis. Strategic guidelines for Menorca, Intitució d'Estudis Menorquins (IME), Govern de les Illes Balears [Online]. Available at: http://www.ime.cat/WebEditor/Pagines/file/Menorca%27s%20first%20energy%20transition.pdf [Accessed 10 October 2018].

Carpintero, Ó. (2005). *El metabolismo de la economía española. Recursos naturales y huella ecológica (1955–2000)*. Fundación César Manrique, Las Palmas.

CIME, C.I. and Ezquiaga, A.S. (2019). Revisión del Plan Territorial Insular de Menorca. Report, Consell Insular de Menorca, Menorca.

Comité español del programa MaB and Organismo autónomo parques nacionales (2017). Plan de Acción de Ordesa-Viñamala 2017–2025 para la Red Española de Reservas de la Biosfera (PAOV-RERB 2017–2025). Report, Ministerio de agricultura y pesca, alimentación y medio ambiente, University of Valencia, Valencia.

Consell Insular de Menorca (2019). Pla d'acció de la reserva de biosfera de Menorca. Report, Reserva de Biosfera de Menorca, 210.

Cooney, R. (2006). A long and winding road? Precaution from principle to practice in biodiversity conservation. *Implementing the Precautionary Principle: Perspectives and Prospects*, 223–244. doi: 10.4337/9781847201676.00022.

Cottrell, W. (1955). Energy and society: The relationship between energy, social change, and economic development [Online]. Available at: https://scholar.google.es/scholar?cluster=9929832534464576676&hl=es&as_sdt=2005&sciodt=0,5 [Accessed 18 March 2019].

Daniels, P.L. and Moore, S. (2002). Approaches for quantifying the metabolism of physical economies: A comparative survey. Part II: Review of individual approaches. *Journal of Industrial Ecology*, 6(1), 65–88. doi: 10.1162/10881980160084042.

d'Estudis, I.M. and CIME, C.I. (2019). Jornades sobre els 25 anys de la reserva de biosfera de Menorca: Conclusions, Revisió del Decàleg RB+20 i Decàleg RB+25. Quaderns de la reserva de biosfera de Menorca, 18, 29 [Online]. Available at: http://jornadesrb.ime.cat/Contingut.aspx?IdPub=14495.

Fischer-Kowalski, M. (1998). Society's metabolism: The intellectual history of material flow analysis, 1860–1970. *Journal of Industrial Ecology*, 2(1), 61–78. doi: 10.1162/jiec.1998.2.1.61.

Fischer-Kowalski, M. and Hüttler, W. (1998). Society's metabolism: The intellectual history of materials flow analysis, 1970–1998. *Journal of Industrial Ecology*, 2(4), 107–136. doi: 10.1162/jiec.1998.2.4.107.

Frame, B. and O'Connor, M. (2011). Integrating valuation and deliberation: The purposes of sustainability assessment. *Environmental Science and Policy*, 14(1), 1–10. doi: 10.1016/j.envsci.2010.10.009.

Fraňková, E., Haas, W., Singh, S.J. (2017). *Socio-Metabolic Perspectives on the Sustainablity of Local Food Systems*. Springer, Berlin. doi: 10.1007/978-3-319-69236-4.

Funtowicz, S.O. and Ravetz, J.R. (1993). Science for the post-normal age. *Futures*, 25(7), 739–755.

Funtowicz, S.O. and Ravetz, J.R. (1994). Emergent complex systems. *Futures*, 26(6), 568–582.

Georgescu-Roegen, N. (1975). Dynamic models and economic growth. *World Development*, 3(11–12), 765–783. doi: 10.1016/0305-750X(75)90079-0.

Giampietro, M. (2019). The quality of scientific advice for policy: Insights from complexity [Online]. Available at: https://magic-nexus.eu/documents/quality-scientific-advice-policy-insights-complexity.

Giampietro, M. (2020). The treacherous use of indicators for SDGs, MAGIC-NEXUS Project [Online]. Available at: https://magic-nexus.eu/content/treacherous-use-indicators-sustainable-development-goals [Accessed 26 July 2020].

Giampietro, M. and Mayumi, K. (2000). Multiple-scale integrated assessment of societal metabolism: Introducing the approach. *Population and Environment*, 22(2), 109–153. doi: 10.1023/A:1026691623300.

Giampietro, M. and Mayumi, K. (2018). Unraveling the complexity of the Jevons paradox: The link between innovation, efficiency, and sustainability. *Frontiers in Energy Research*, 6, 26. doi: 10.3389/fenrg.2018.00026.

Giampietro, M., Mayumi, K., Ramos-Martin, J. (2008). Multi-scale integrated analysis of societal and ecosystem metabolism (MuSIASEM): Theoretical concepts and basic rationale. *Energy*, 34(34), 313–322. doi: 10.1016/j.energy.2008.07.020.

Giampietro, M., Aspinall, R.J., Ramos-Martin J., Bukkens, S.G.F. (2014). *Resource Accounting for Sustainability Assessment: The Nexus between Energy, Food, Water and Land Use*. Routledge, London.

Ginard, X. and Murray, I. (2015). El metabolismo socioeconómico de las Islas Baleares, 1996–2010. In *El metabolismo económico regional español*, Carpintero, Ó. (ed.). FUHEM Ecos, Madrid.

González de Molina, M. and Toledo, V.M. (2014). *The Social Metabolism*. Springer, Berlin. doi: 10.1007/978-3-319-06358-4.

IDS, I. and d'Estudis, I.M. (1996). Plan de Desarrollo Sostenible. Estudio de viabilidad. Consell Insular de Menorca (CIME). United Nations Educational, Scientific and Cultural Organization (UNESCO) and LIFE Project (European Commission), 154.

Koestler, A. (1967). *The Ghost in the Machine*. Arcana Books, Culver. doi: 10.1186/2041-2223-1-11.

Koestler, A. (1979). Janus: A summing up. *Bulletin of the Atomic Scientists*, 35(3), 4–4. doi: 10.1080/00963402.1979.11458590.

Kovacic, Z. (2015). Complexity theory in quality assessment. PhD Thesis, Autonomous University of Barcelona, Barcelona.

Kovacic, Z. and Giampietro, M. (2015). Beyond GDP indicators: The need for reflexivity in science for governance. *Ecological Complexity*, 21, 53–61. doi: 10.1016/j.ecocom.2014.11.007.

Lotka, A.J. (1922). Natural selection as a physical principle. *Proceedings of the National Academy of Sciences*, 8(6), 151–154. doi: 10.1073/pnas.8.6.151.

Lotka, A.J. (1956). *Elements of Mathematical Biology*. Dover Publications, New York.

Marcos-Valls, A., Kovacic, Z., Giampietro, M., Kallis, G., Rieradevall, J. (2020). Isolated yet open: A metabolic analysis of Menorca. *Science of the Total Environment*, 738, 139221. doi: 10.1016/j.scitotenv.2020.139221.

Martinez-Alier, J. (1987). *Ecological Economics: Energy, Environment and Society*. Blackwell, Oxford.

Martinez-Alier, J., Kallis, G., Veuthey, S., Walter, M., Temper, L. (2010). Social metabolism, ecological distribution conflicts, and valuation languages. *Ecological Economics*, 70(2), 153–158. doi: 10.1016/j.ecolecon.2010.09.024.

Martínez-Iglesias, C., Sorman, A.H., Giampietro, M., Ramos-Martin, J. (2014). Assessing biophysical limits to the economic development of remote islands: The case of Isabela in the Galapagos Archipelago. Working paper, Instituto de Altos Estudios Nacionales, Centro de Prospectiva Estratégica, 1, 1–26. doi: 10.13140/RG.2.1.1518.7924.

Maxim, L. and van der Sluijs, J.P. (2007). Uncertainty: Cause or effect of stakeholders' debates? Analysis of a case study: The risk for honeybees of the insecticide Gaucho. *Science of the Total Environment*, 376(1–3), 1–17. doi: 10.1016/j.scitotenv.2006.12.052.

Menton, M., Larrea, C., Latorre, S., Martinez-Alier, J., Peck, M., Temper, L., Walter, M. (2020). Environmental justice and the SDGs: From synergies to gaps and contradictions. *Sustainability Science*, 15, 1621–1636. doi: 10.1007/s11625-020-00789-8.

Obsam, O.S. (2020). Informes d'indicadors [Online]. Available at: https://www.obsam.cat/informes-indicadors/.

O'Connor, M. (2006). The "Four Spheres" framework for sustainability. *Ecological Complexity*, 3(4), 285–292. doi: 10.1016/j.ecocom.2007.02.002.

Ostwald, W. (1907). The modern theory of energetics. *The Monist*, 481–515.

Ostwald, W. (1911). Efficiency. *The Independent*, 71, 867–871.

Pérez-Sánchez, L., Giampietro, M., Velasco-Fernández, R., Ripa, M. (2019). Characterizing the metabolic pattern of urban systems using MuSIASEM: The case of Barcelona. *Energy Policy*, 124, 13–22. doi: 10.1016/j.enpol.2018.09.028.

Singh, S.J., Grünbühel, C.M., Schandl, H., Schulz, N. (2001). Social metabolism and labour in a local context: Changing environmental relations on Trinket Island. *Population and Environment*, 23(1), 71–104. doi: 10.1023/A:1017564309651.

Soddy, F. (1926). *Virtual Wealth and Debt: The Solution of the Economic Paradox*. Britons Publishing Company, London.

Spaiser, V., Ranganathan, S., Swain, R.B., Sumpter, D.J.T. (2017). The sustainable development oxymoron: Quantifying and modelling the incompatibility of sustainable development goals. *International Journal of Sustainable Development and World Ecology*, 24(6), 457–470. doi: 10.1080/13504509.2016.1235624.

Swyngedouw, E. (2006). Circulations and metabolisms: (Hybrid) Natures and (Cyborg) cities. *Science as Culture*, 15(2), 105–121. doi: 10.1080/09505430600707970.

UNESCO (1993). Menorca Biosphere Reserve (Spain). Report, UNESCO.

UNESCO (1996). The seville strategy and the statutory framework of the world network. Report, UNESCO, 18.

UNESCO (2014). Agreement between the United Nations Educational, Scientific and Cultural Organization (Unesco) and the Kingdom of Spain regarding the establishment of the "International Centre on Mediterranean Biosphere Reserves, two coastlines united by their culture and nature" as a category 2 centre under the auspices of UNESCO. Report, UNESCO.

UNESCO (2016). Lima action plan for Unesco's Man and the Biosphere (MAB) Programme and its World Network of Biosphere Reserves (2016–2025) (March) [Online]. Available at: http://www.unesco.org/fileadmin/MULTIMEDIA/HQ/SC/pdf/Lima_Action_Plan_en_final.pdf.

UNESCO (2017). MAB Strategy (2015–2025), Lima Action Plan (2016–2025) and Lima declaration. A new roadmap for the Man and the Biosphere (MAB) Programme and its World Network of Biosphere Reserves [Online]. Available at: https://unesdoc.unesco.org/ark:/48223/pf0000247418.

UNESCO (2020). Man and the Biosphere (MAB) Programme [Online]. Available at: https://en.unesco.org/mab [Accessed 21 August 2020].

United Nations (2015). Transforming our world: The 2030 agenda for sustainable development, A/RES/70/1. Report. doi: 10.1163/157180910X12665776638740.

Velasco-Fernández, R. (2017). The pattern of socio-ecological systems. A focus on Energy, Human Activity, Value Added and Material Products. PhD Thesis, Autonomous University of Barcelona, Barcelona.

Velasco-Fernández, R. (2020). Depoliticizing and repoliticizing SDGs MAGIC-NEXUS Project [Online]. Available at: https://magic-nexus.eu/content/depoliticizing-and-repoliticizing-sdgs [Accessed 2 July 2020].

Velasco-Fernández, R., Ramos-Martín, J., Giampietro, M. (2015). The energy metabolism of China and India between 1971 and 2010: Studying the bifurcation. *Renewable and Sustainable Energy Reviews*, 41(1). doi: 10.1016/j.rser.2014.08.065.

Weitz, N., Carlsen, H., Nilsson, M., Skånberg, K. (2018). Towards systemic and contextual priority setting for implementing the 2030 agenda. *Sustainability Science*, 13(2), 531–548. doi: 10.1007/s11625-017-0470-0.

White, L.A. (1943). Energy and evolution of culture. *American Anthropologist*, 45(3), 335–356.

Wolman, A. (1965). The metabolism of cities. *Scientific American*, 213, 156–174. doi: 10.1017/CBO9781107415324.004.

Zipf, G.K. (1941). *National Unity and Disunity. The Nation as a Bio-social Organism.* Principia Press, Bloomington.

6

The Jabal Moussa Biosphere Reserve (Lebanon): A Private Association Initiative

6.1. Introduction

Country	Lebanon
Name of the biosphere reserve	Jabal Moussa
Date of UNESCO designation	2009
Management structure	Association for the Protection of Jabal Moussa (APJM)
Area	6,500 ha Core area: 1,250 ha
Population	Approximately 8,500 ha

Table 6.1. *Profile of the Jabal Moussa Biosphere Reserve*

6.2. Rich by nature

The Jabal Moussa massif is a grand natural space, ideal for walks and hikes in landscapes not often seen in Lebanon. The summit is particularly beautiful, with its dolines and jagged crests, and its large groupings of hop-hornbeam, maple and ash, its great Turkey and Aleppo oaks and its sturdy Syrian junipers, with its glades and ancient terraces. The northern slope has its own stunning surprises, woods of Turkish pine, ancient coal deposits, sheer cliff-faces, a stream surrounded by plane trees [...] and a walk from Ebré to the easternmost point of the massif is a highly agreeable experience.

Chapter written by Pierre DOUMET and Joelle BARAKAT.

This is the conclusion of a detailed report from the French National Office for Forests, funded by the European Union on behalf of the Lebanese Ministry of Agriculture at the end of the 1990s.

Figure 6.1. *Landscape of the Jabal Moussa Biosphere Reserve on the Adonis River*

In 2006, it was accorded the status of protected forest by the Ministry of Agriculture, at the request of the Association for the Protection of Jabal Moussa (APJM) in 2006, followed by its UNESCO designation as a biosphere reserve in 2009.

The impact of climate change on the forest and its species, as well the socio-economic quantification of goods and services provided within the protected area, have been studied within a Mediterranean FFEM (French Funding for the Global Environment) project (Daly Hassen 2016), for which the reserve was a pilot site in Lebanon. Once this quantification was carried out, the project proceeded to research means of optimizing the goods and services offered within the new administrative context (ecotourism, the training of women to sell local products, apiculture, etc.). A study by Poyatos et al. (2016), undertaken in collaboration with the APJM and the Lebanese Ministry of Agriculture, has highlighted the positive impact of the protective measures put in place by the APJM since 2007, noting an increase in dense areas of forest as well as a change in the NDVI index that can be taken as marking an evolution in the health and density of plant coverage.

Research on the region's resident and migratory birds (Demopoulos 2008) has concluded, based on a field study that took place over multiple months, that Jabal Moussa is worthy of being designated a global Important Bird Area (according to the criteria of BirdLife International).

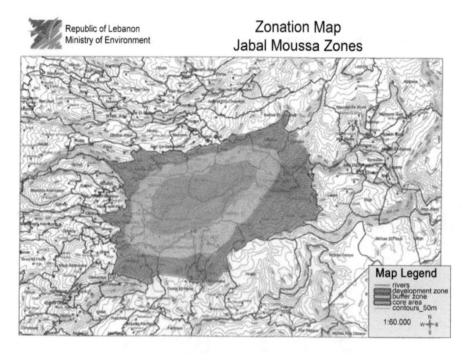

Figure 6.2. *Zoning of the Jabal Moussa Biosphere Reserve. For a color version of this figure, see www.iste.co.uk/romagny/biosphere2.zip*

Figure 6.3. *Long-legged buzzard*

A mammal study by Abi Said and Amr (2012) revealed the "peaceful" presence (visible during the day, rather than exclusively at night) of emblematic mammals such as the wolf and the hyena, as well as 21 other mammals in sufficient quantities as to arouse some hope for the repopulating of natural environmental reserves.

A field expedition by Finnish entomologists (Kurina et al. 2015; Zeller et al. 2016) discovered two insects native only to Jabal Moussa (one of which being *Micropterix jabalmoussae*).

The famous twin figureheads of Lebanese biology/ecology, Georges and Henriette Tohmé, published in 2012 the fascicules *Fleurs* and *Arbres du Jabal Moussa* (*Flowers* and *Trees of Jabal Moussa*), revealing the presence of 726 types of plants at Jabal Moussa, 24 being endemic to Lebanon and six to the site itself. A recent study, carried out within the context of an international project relying on the CEPF (Critical Ecosystem Partnership Fund/AFD), two Lebanese universities (USJ, LAU) and the Jabal Moussa BR, illustrates more closely the interaction between locals, visitors and researchers.

Figure 6.4. *An example of the exceptional plant Pentapera sicula libanotica (source: Tohmé and Tohmé (2012))*

6.3. A privately run biosphere reserve

The governance of the biosphere reserve is the responsibility of the Association for the Protection of Jabal Moussa, a non-profit NGO made up, at the level of its leadership council, of local and national figures, reliant on a young and dynamic local team, in charge of numerous conservation and sustainable development projects funded by national and often international donors. Since the reserve's ownership is split between multiple municipalities (as communal properties), religious institutions (WAQF) and private owners, it would have been natural to involve these parties in the management process in an organized framework. However, given the disparate interests of these parties, it was proven more practical for the APJM to engage with each owner in a system of land-lease contracts specifying that the property be used in accordance with the objectives of the MAB program. The mayors and other local elected officials, the religious representatives and some private owners have thus become sporadic participants who have, in a sense, given the APJM proxy power to conscientiously manage their properties according to the principles of the MAB program/UNESCO.

It is important to note that the Jabal Moussa enjoyed no protection before the founding of the APJM. Given the numerous levels of protection (protected forest, natural site, archaeological sites obtained by intermediaries of the APJM), the relevant ministries (Agriculture, Environment, Culture) should theoretically also have become part of the governing body of the reserve.

In practice, contacts with these bodies are themselves sporadic. Originally asked by the APJM to lend their protection to the site, these institutions are now dealing with a variety of other preoccupations and verify, during visits for festive events or the beginnings and ends of projects, that work is being carried out correctly and development is continuing with regard to natural and cultural heritage.

The APJM is made up of an assembly of members (meeting annually), a leadership council (meeting quarterly) and a management structure which oversees the day-to-day running of the reserve.

After two years of working alongside the different institutions, in 2012, the APJM presented its first 10-year plan for sustainable management (supported by the Swiss MAVA foundation and UNESCO Beirut) during an interactive conference involving academics, researchers, local officials, religious representatives and local communities from within the reserve.

6.4. International recognition

The APJM has accumulated international designations for the site, from UNESCO (biosphere reserve), BirdLife International (Global IBA), the IUCN (as an example of private protected area "best practice"), from MedMAB, as well as the Lebanese protection labels of "protected forest" (from the Ministry of Agriculture) and "protected archeological site" (Ministry of Culture).

Figure 6.5. *Archaeological dig in a tomb*

6.5. Administration led by socio-economic expectations

The major axes of administration concern biodiversity, the understanding of cultural heritage, especially archaeology, academic research and above all socio-economic development. They were subject to constant evolution and a continual dialogue with the parties engaged in the initial 10-year plan.

Figure 6.6. *Promotion of local products*

Two polls have been carried out, with an interval of 10 years, in 2009 and 2020 (Roula Abi Habib Khoury forthcoming), to analyze the impact of 10 years of work by the APJM in the principal villages of the reserve and to evaluate the perception of inhabitants of villages in and around the core area, with regard to the update to the 10-year management and development plan expected for 2022. The comparison shows that the biosphere reserve is now considered to be a major employer and that the feeling of "liberties being trampled on" is now less acute than it was at the beginning. The idea of ecotourism, with its increase in the annual number of visitors from 300 to 30,000 (between 2009 and 2019), is now well understood, and for the most part accepted. This is due to the number of jobs created: employees; guides; guards; jobs in guest houses; or working with local products. Young people (as employees or guides) and women (working in guest houses and local products) make up the majority of beneficiaries. A part of the population continues to demand unlimited access to all activities, licit or illicit, while another fears an increase in savage animals. Let us note that in 2020, the year of the start of the Covid-19 pandemic, the number of visitors rose a little above 40,000, with natural spaces frequented by many families.

Figure 6.7. *Visitors are attracted by the many walks, such as this roman staircase*

All of the meetings held through the years with mayors and the inhabitants of their villages support the following statement: in a context of general impoverishment and an expanding economic crisis, the foremost demand is for direct socio-economic benefit. Local participation can thus be summed up, for the most part, as a desire to orient the windfall of ecotourism and the sale of products towards their own village.

Beekeeping and the gathering and preparation of thyme are increasingly popular among locals. These products (sold under the Jabal Moussa label) are distributed at a national level and are growing in recognition. The standards of production are extremely strict and the preparation of these products (as well as others, such as jams and artisan crafts) takes place in a well-equipped central workshop/kitchen.

The plants produced by Jabal Moussa's nursery, the seeds of which come from the biosphere reserve, have a strong reputation, according to the two primary national bodies for reforesting, and are thus highly prized. The APJM is one of the newest Lebanese participants in the CNTPL (Cooperative of Native Tree Producers of Lebanon). The reforesting of the interior of the reserve is selective and progressive.

6.6. Efforts at increasing understanding and awareness of an exceptional biodiversity

Jabal Moussa and its environs have been designated a Key Biodiversity Area (KBA) by the Critical Ecosystem Partnership Fund (CEPF), including the results of an investigation by BirdLife International. The CEPF's national study showed that the Jabal Moussa/Adonis Valley KBA has the highest score in Lebanon with regard to Strategic Direction 4, which focused on the plants that are in critical danger of extinction or have very narrow distribution areas.

A project to conserve flora endemic to Lebanon through community engagement is financed by the CEPF, and involves the University of Saint-Joseph (USJ), the NGO Friends of Nature and the APJM. It regards the conservation of the flora endemic to Lebanon, among which is that of the Jabal Moussa Biosphere Reserve, known for its endemic species such as *Cyclamen libanoticum, Erica sicula libanotica, Salvia peyronii, Paeonia kesouanensis*, etc.

The project's goal is to raise awareness of species endemic, indigenous and unique to Jabal Moussa and its environs. Some activities have been carried out in collaboration with Friends of Nature and the University of Saint-Joseph, such as field evaluations of the status of species and the botanical training of members of the APJM and the production of a management plan to better conserve the reserve's flora, as well as the development of awareness tools for the community in the form of pedagogical sites: herbariums, botanical gardens, mobile applications, social media campaigns, etc.

The research led by the Jabal Moussa Biosphere Reserve has two poles:

– conservation in situ, which involves identifying and mapping, with the help of locals and visitors, the floral species endemic to Jabal Moussa, and evaluating their status so as to establish conservation actions;

– conservation ex situ, through the elaboration of germination protocols for endemic species. Once established, these protocols should allow the APJM to propagate these endemic species in the native species nurseries.

The APJM must relay these facts and results to the public, through awareness activities for local communities and visitors to the reserve:

– a welcome center ("The Budding Botanist") exists for schoolchildren and those interested in botany, including an exhibition on particular species of flower, an herbarium focused on the reserve, and an activity room;

– this center is close to three native plant nurseries with which it enjoys close ties: visitors can observe the process of propagating native plants from seeds, and participate in planting activities;

– a space called "the biosphere garden" at the entrance to the reserve, where visitors can walk, relax and deepen their botanical knowledge by observing the species present in the garden, whether it be native species introduced by the APJM (e.g. a series of cyclamens including the endemic *Cyclamen libanoticum*), so as to learn how to distinguish the native species, or whether it be a zaatar population (*Origanum syriacum*, a culinary herb highly prized both locally and at the national level);

– a botanical walk outfitted with panels bearing the name of the plant and a QR code leading to a mobile application providing information on the plant;

– a social media awareness campaign on the importance of endemic species.

6.7. References

Abi-Said, M.R. and Amr, Z. (2012). Camera trapping in assessing diversity of mammals in Jabal Moussa Biosphere Reserve, Lebanon. *Vertebrate Zoology*, 62(1), 145–152.

Daly Hassen, H. (2016). Assessment of the socio-economic value of the goods and services provided by Mediterranean forest ecosystems: Critical and comparative analysis of studies conducted in Algeria, Lebanon, Morocco, Tunisia and Turkey. Report, Blue plan, Valbonne.

Demopoulos, H. (2008). A study into the importance of Jabal Moussa for birds in Lebanon – A Rocha Lebanon [Online]. Available at: https://lebanon.arocha.org/wp-content/uploads/sites/17/2015/04/ARL-SC-08-4-Jabal-Moussa-Report-A-STUDY-INTO-THE-IMPORTANCE-OF-JABAL-MOUSSA-FOR-BIRDS-IN-LEBANON.pdf.

Khoury Roula, A.H. (2011). Enquête socioéconomique dans la région de Jabal moussa. Communautés et société, annales de sociologie et d'anthropologie. PhD Thesis, Université Saint-Joseph, Beyrouth.

Kurina, O., Õunap, E., Põldmaa, K. (2015). Two new Neuratelia Rondani (Diptera, Mycetophilidae) species from Western Palaearctic: A case of limited congruence between morphology and DNA sequence data. *ZooKeys*, 496, 105–129.

Poyatos, M., Lara-Gómez, M., Varo, M., Navarro Cerrillo, R., Palacios, G. (2016). Vulnerability assessment of Mediterranean forest ecosystem to climate change impacts on Quercus ceris L. var. pseudo-cerris (Boiss.) Boiss Quercus calliprinos Webb. and Pinus brutia Ten. Populations [Online]. Available at: http://ffem-project.idaf.es/files/report.pdf.

Tohmé, G. and Tohmé, H. (2012). "Fleurs" et "Arbres" du Jabal Moussa. Report, Université St Joseph Beyrouth.

Zeller, H., Kullberg, J., Kurz, M. (2016). A new species of *Micropterix* Hübner, 1825 from Lebanon (Lepidoptera: Micropterigidae). *Nota Lepidopterologica*, 39(2), 101–107.

7

Understandings of Administration and Challenges to Governance in the Arganeraie Biosphere Reserve (Morocco)

7.1. Introduction

Country	Morocco
Name of the biosphere reserve	Arganeraie
Date of UNESCO designation	1998
Management structure	Regional Directorate for Water and Forests
Areas	2.5 million ha
Population	3 million ha

Table 7.1. *Profile of the Arganeraie Biosphere Reserve*

7.2. A biosphere reserve built around an iconic tree: the argan tree

Situated in the south-west of Morocco, the Arganeraie Biosphere Reserve is a group of plains bordered by the High Atlas and Anti-Atlas mountains, and by the Atlantic Ocean to the west. Its territory ranges in altitude from sea level to 2,500 meters. It is part of the Mediterranean/Saharan transition zone, established around a species of tree endemic to Morocco, the argan tree (*Argania spinosa*), which dates back to the tertiary era (between 65 million and 1.8 million years ago). As typical of the Macaronesian area, the Arganeraie is home to sclerophyllous Mediterranean

---

Chapter written by Abdelaziz AFKER.

forests, woodlands and scrub. It obtained its UNESCO Biosphere Reserve designation with the aim of sustainably managing its socio-economic and ecological systems, while ensuring the sustainable development of the local and regional economy.

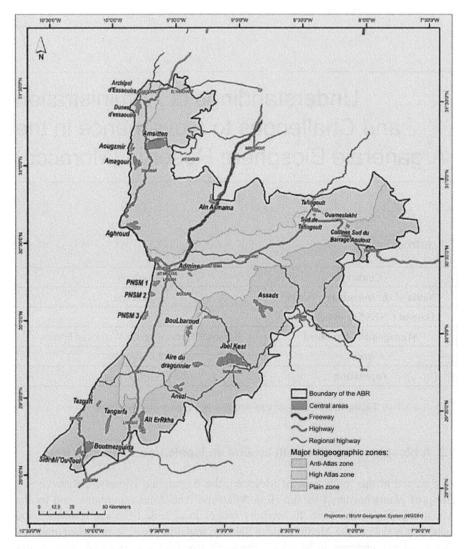

Figure 7.1. *Zoning of the Arganeraie Biosphere Reserve. For a color version of this figure, see www.iste.co.uk/romagny/biosphere2.zip*

Figure 7.2. *The argan tree, a treasure of the biosphere reserve (source: A. Afker)*

The major challenge of the Arganeraie Biosphere Reserve is the lack of resilience of its ecosystems, a lack closely tied to socio-economic factors, economic opportunities and social changes molded by the dynamics of the area. We can nonetheless point out a positive: the local population has now become aware of the economic and ecological role of the argan tree.

7.3. An integrated approach to conservation and ecodevelopment

The status of biodiversity is highly varied, with some areas subject to strong anthropic pressure and others showing positive changes. Thus, the spectacular return of the Cuvier gazelle in the Anti-Atlas mountains, previously highly endangered, is an indicator of the improvement of some natural habitats. Today, however, the species is regularly the target of poachers. Local administrations have become very active with regard to its preservation and spare no effort to support surveillance and protection tasks.

The Agency for Water and Forests, the role of which is to protect biodiversity – particularly wild flora and fauna – makes use of regional preservation plans. They are elaborated on the basis of the state, the threats to and the status of species, considered at the national and international level (IUCN).

The biosphere reserve offers spaces classified as national parks and 14 Sites of Biological and Ecological Interest (SIBEs), 12 of which are on land.

Figure 7.3. *Pastoralism in the Arganeraie Biosphere Reserve (source: A. Afker)*

Different actions are taken that directly or indirectly serve to preserve biodiversity:

– assessments of the biodiversity status of several forested and peri-woodland massifs, drainage basins and protected areas;

– rehabilitation activities and activities aimed at reconstituting degraded spaces and ecosystems;

– conservation activities centered on species that are either endangered or of serious ecological value;

– the reintroduction of endangered or extinct animal species;

– measures to protect soil against water erosion and desertification.

It is important to note that the role of the population in the targeting of these activities has become essential. It has considerably changed over these last few years, and all biodiversity conservation programs and the regeneration of indigenous species (particularly the argan tree) rely on negotiations around the forests to be chosen. Negotiation and compromise are keenly sought before and during the implementation of these projects. The creation of synergies between the actors is motivational, as is the idea of compensation for rights of use. In addition to this, an average of 5,000 ha of regenerated argan trees and other indigenous species is achieved each year within the Arganeraie Biosphere Reserve. The most operational formula consists of pairing interventions with revenue-generating activities, such as the distribution of apiaries: local and national markets demand great quantities of honey from the biosphere reserve, particularly thyme and spurge honey, highly sought after as a natural remedy and health product. This approach has allowed for the creation of a beekeeping industry, in addition to the existing large-scale beekeepers. A great number of people and a large part of the Arganeraie Biosphere Reserve are involved.

There exists an ecotourism project involving one of the core areas, the Souss Massa National Park, where two associations have been set up to offer services to visitors: an organization of hiking guides, and an association of observation guides at the mouth of the Massa. Their members, all local, have benefited from training and logistical support. An ornithological circuit, a loop of 5 km allowing for the observation of water birds, and a nature circuit of 4 km, which can be followed on foot or on a donkey, have also been created.

At the center of the same national park, a women's cooperative has been created for the promotion of mussels. The products are now commercialized according to quality standards certified by qualified organizations. In moving from a traditional, informal mode to a professional one, their revenue has increased fivefold.

Figure 7.4. *Traditional home within the Arganeraie Biosphere Reserve (source: A. Afker)*

A rural tourism development project has been established, centered on promoting the reserve's hinterlands: it aims to establish a network of ecological niches, allowing for the promotion of their products and the proposal of reception services. Reception services thus create relationships between visitors, the women's cooperative and small agricultural producers (of fresh vegetables, honey, etc.).

7.4. Participation-oriented administration

The administration of the Arganeraie Biosphere Reserve relies on two bodies:

– A managing body, the Regional Directorate of Water and Forests, based in Agadir. It ensures coordination between the provincial, regional and national levels, within the framework of its regular activities. Attached to this body are regional and local programs tied to the conservation and promotion of biodiversity. This directorate shapes the area's dynamics in favor of the reserve, while aiming to adhere to its three statutory functions.

– A larger participative body, which is made up of actors tied to the area of the biosphere reserve. Thus, the Regional Directorate of Water and Forests works regularly and closely with different key actors, such as the Agency for the Development of the Arganeraie, and the Federation of Rights Holders and the Regional Coordination of the Network of Associations of the Arganeraie Biosphere Reserve (RARBA). They support the managing body's coordination and stimulation efforts in a concerted and complementary manner, contributing to the meeting of its objectives.

The National Federation of Rights Holders and Users of the Arganeraie (FNADUA) gathers the provincial associations and rights holders of the Arganeraie across eight provinces. This representational body has become a legitimate and essential interlocutor for the institution's representative bodies.

The regional coordination of RARBA is composed of provincial coordinating bodies. It gathers a number of local associations tied to villages or groups of villages. This network is today the key actor speaking for the Arganeraie Biosphere Reserve and the regional and national level. The RARBA has recently supported the creation of an association of women active in associations and politics.

In the Arganeraie Biosphere Reserve's never-ending search for participative and inclusive management, institutions and representatives of the population have become the primary categories of stakeholders, ensuring a greater representation of interest groups of varying levels of involvement. Associated are:

– The Moroccan Interprofessional Federation for the Argan Industry (FIMARGANE), officially recognized as a key actor in the sustainable development of the argan product industry. Rights holders, who are owners or providers of the raw material, hold a privileged place in this organization. Women are equally present through the union of (women-only) cooperatives for the extraction of argan oil.

– The regional arm of the Moroccan Alliance for the Climate and for Sustainable Development (AMCDD) is also active in advocacy for sustainable development issues. It deals with issues related to Sustainable Development Goals (SDGs). The RARBA, connected with this organization, can thus link SDGs to the stimulation of the Arganeraie Biosphere Reserve.

Rights holders stand as the central actors in the practical governance of the biosphere reserve, thanks to their power to use the Arganeraie's resources, in particular its fruit, labor and pastures. They can thus be seen as local actors with a strong connection with the space, its culture and its identity.

The Arganeraie Biosphere Reserve is recognized as a space optimized for agroforestry and pastoral functions. Three production systems are present at various scales, with its numerous interest groups and interdependences subject to the whims of the climate. However, all of these systems have seen both spatial and temporal changes, following changes to their socio-economic and religious contexts. Thus, conflicts of interest and use have arisen to condition and impact the processes of negotiation and involvement of the various categories of actors.

The reserve is thus home to a complex system of open spaces inextricable from usage rights, and to many actors forming different interest groups. The management of its assets is a true ambition, both on the part of multi-sector managers and of mere users. It is highly complex, and a synthesis of ambitions and means is keenly sought. In such a context, the search for compromise within these multi-sector and multi-actor dynamics requires a great deal of effort and time.

In terms of management, an action plan for the Arganeraie Biosphere Reserve (2018–2027) has been created following the reserve's second 10-year periodical review.

The management model best suited to the reserve's particularities and challenges, though, is still subject to debate and reflection, weighing the benefit of its territories, on the one hand, and the attainment of an acceptable and sustainable conciliation threshold for its functions, on the other hand. This debate touches on governance, and on the denial of the concept of the biosphere by acts on the ground. There are still divergences, some considering the biosphere reserve to be a label and others seeing it as a land management project ensuring the implementation of collective responsibility, or indeed as a concept to be applied in any way possible on the ground.

7.5. Regarding the research/education/management dialogue

Scientific research has been a part of the biosphere reserve from the very beginning. Universities and research institutions were already interested in the argan tree and its natural, cultural, social, economic and heritage value. However, since 2011, the Argan International Congress, organized biennially by the National Agency for the Development of Oasis and Argan Zones (ANDZOA) and its partners, has strongly encouraged such research. The Congress is a platform for the presentation of research, achievements and projects carried out within the biosphere reserve. The workshops and round tables offer opportunities for discussions on challenges debated at the national and international level, notably around adaptation to climate change, sustainable development and all areas linked to sociology, anthropology, etc.

A recommendation from the Congress – the creation of a platform for dialogue and cooperation to benefit the Arganeraie Biosphere Reserve – has just come to life with the creation of the National Argan Center: it aims to develop scientific research in dialogue with the needs of administrators and the needs of the land. It should be noted that there are notable differences between the academic research led by universities and the applied research carried out by private-industry institutions to meet generally technical and highly specific needs. However, the last Congresses have allowed for very interesting debates between the scientific community and socio-professional organizations, and between civil society and elected officials. They have demonstrated the value of scientific research for the biosphere reserve, and the place of knowledge on management and on the reserve's day-to-day practices.

In conclusion, the scientific research/management dialogue is unavoidable for the Arganeraie Biosphere Reserve. It can only productively and actively develop its land through the involvement of local populations in the process of defining problematics and possible projections for the future.

7.6. References

Aboutayeb, H. (2015). *Tourisme durable dans la reserve de biosphère de l'Arganeraie au Maroc. Plaidoyer pour l'implémentation d'une stratégie de développement touristique durable.* Éditions Universitaires Européennes, Saarbrücken.

Faouzi, H. and Martin, J. (2014). Soutenabilité de l'arganeraie marocaine. *Confins*, 20.

8

Reconciling Conservation and Sustainable Development: The Example of the Arganeraie

8.1. Introduction

The Arganeraie Biosphere Reserve (ABR), highly extended and situated in the center-west of Morocco, is an open-air laboratory for the tracking of different types of transformations and changes. These are the product of interactions between different factors linked to economic, social and cultural innovations, to adaptations, to the effects of globalization and to different environmental changes. The ABR's trajectory can teach us about the complexity of governing its lands, about how to respond to the juxtaposed and sometimes opposed challenges of its conservation and its sustainable territorial development. The Arganeraie Biosphere Reserve's UNESCO designation is now being put to the test by these temporal and spatial evolutions, at the same time as it is carrying out its second periodical review. Once a decade, a review of its UNESCO recognition is carried out, investigating the new realities of the reserve. A report was produced for the period 1998–2008, the reserve's UNESCO designation being maintained following its publication. In 2018, a new periodical review report was produced with a wider range of elements and parameters, compared to the highly simplified standard model used in the first evaluation.

What lessons can be taken from this second evaluation of the ABR? What changes have occurred in conservation and sustainable territorial development

Chapter written by Abdelaziz AFKER and Saïd BOUJROUF.

within the context of climate change and the promotion of local particularities? How can differing public policies regarding deconcentration and decentralization, the implementation of SDGs, and the different scales of public actions be reconciled and expanded? Should the ABR be considered a regional project, or simply a brand or label? How should conservation and development be balanced?

The Statutory Framework for the World Network (UNESCO 1996) demands that the state of each biosphere reserve be made the object of a periodical review every 10 years, on the basis of a report that the authority in charge will establish, and that the state concerned will address to UNESCO's MAB secretariat. This examination will aim to verify that the site is responding to the criteria (article 4 of the Statutory Framework) which allowed for its designation.

With time and practice, this periodical review has become in multiple countries a key phase in the adaptive management of biosphere reserves: it is an opportunity to take stock of the actions which have been carried out, of the implementation of management policies, of governance, and to evaluate successes and challenges and consider them within the perspective of the next 10 years of management. This opportunity has also been used to reengage stakeholders in the region, and to encourage local participation.

Box 8.1. *Periodical reviews of biosphere reserves*[1]

8.2. The ABR, between conservation and sustainable territorial development: reconciling the irreconcilable

Today, argan habitats in Morocco are undergoing a clear and profound transformation, a paradigm shift. They are moving with some difficulty towards an adoption of UNESCO's biosphere reserve concept. This new system of reference calls for a general renewal of basic knowledge, of the concepts used to best frame and translate the changes currently being experienced into scientific representations and interpretations.

The ABR is facing, now more than ever, a powerful degradation of its soils, drainage basins, biodiversity and hydrographic potential. The effects of human actions are visible across the reserve. Changes in climate and changes across the globe are particular stressors. Even though "the specific values of biosphere reserves and the opportunities they represent with regard to climate change have been recognized within the Madrid Action Plan for Biosphere Reserves (2008–2013) and the Dresden Declaration on Biosphere Reserves and Climate Change (2011)"

---

1 Written by C. Cibien.

(UNESCO 2017), this recognition leaves the ABR to confront more complex and increasingly broad environmental changes without any effective actions. The involvement of different actors and particularly the inclusion of the local population could contribute to the resilience of these areas, allowing for a reinforcing of "the conservation and sustainable management of biodiversity which forms an integral part of sustainable development" (OECD 1998). To this end, we will move away from strict conservation, and from the condemnation of local rural practices, to the integration of "local communities" in the conservation of spaces and species (Wells and Bradon 1992; Martin 2002) so as to ensure, as far as possible, sustainable territorial development.

Actors are engaged through the launching of promotional and patrimonializational processes, mobilizing specific resources which could act as the basis of a general label such as that endowed by UNESCO in 1998, or of a more specific label like the AOP-IGP for argan oil, following a 2008 law.

Promotion, above all touristic, espouses the patrimonialization of touristic villages such as Agadir. This relationship between promotion and the recognition of the biosphere reserve at an international level is a strong problem (Gravari-Barbas et al. 2012), the result of economic consequences and of a real "labeling race" (Duval and Smith Benjamin 2014). In fact, labeling has become a means of justifying the discourses and practices of environmental protection (Tommasi et al. 2017), as well as of reinforcing connections to the land, and the reconstruction of resources (François et al. 2013) which become not only connected to that land, but increasingly viewable as heritage (Boujrouf 2014).

While general or specific labeling contributes to the construction of a regional brand and to a local identity, specification systems become increasingly banal, and risk eroding the brands and identities they have constructed. The idea that labels for regional products protect local knowledge needs to be demonstrated (Boisvert and Caron 2010), especially given the strong deviations noted in different places and regions within the ABR.

These trends increasingly intersect with the profound transformations that are creating a duality, opposing traditional and modern regions by provoking waves of emigration from marginal areas and great resistance from impoverished corners (El Fasskaoui 2009; Faouzi 2017). For this reason, these juxtapositions produce, in an evolving sense, new social figures deconstructing family farming (Lacombe 2015) at a time when we should be moving towards innovations and resilience that would allow for the return of agroecological sustainability, or indeed sustainable territorial development.

8.3. The complex challenges characterizing the ABR, or relevance and adaptation in conciliatory resilience

The RBA's vast territory is characterized by its strong contrasts, and by the fragile equilibria of its dominant arid and semi-arid landscapes, lacking in resilience. Climatic aspects, in particular rain levels and temperatures, are also highly varied, as are socio-economic variabilities and vulnerabilities. This expansive territory also offers diversified heritage capital, linked to culture and to local know-how.

Erosion varies greatly according to the resource in question and its uses. One important area is threatened by water erosion, linked to irregular rains, the lack of density of its vegetation and the preponderance of rugged terrain. Wind erosion is also a strong threat, thanks to the dominance of sandy substrates and the influence of winds in certain areas. These effects are aggravated locally by over-grazing, and the nature of and transformations to the use of adapted cultural techniques. The irregularity and weakness of seasonal rains leads to resource scarcity with regard to surface water. The dominant form of agriculture, which due to space is largely based around grains, contributes marginally to revenue (bour) and remains subject to climatic conditions; the forest ecosystems compensate for this in times of scarcity. Furthermore, the important spatiotemporal differences (upstream/downstream and east/west) in the dynamics of local development are also significant.

The lack of resilience of these areas is coupled with the large-scale effects of socio-economic conditions (poverty), with the dynamic imposed by the rapid local development of inhabited spaces and with the various factors in the degradation of ecosystems, demonstrating a vulnerability which makes natural threats imminent.

At the same time, the physical landscape and the typology of ecosystems, combined with the qualities of uses, leads to a kind of stratification which allows for the identification of homogeneous territories where particular interventions could be implemented. This stratification is more refined when it takes into account the achievements of prior experiences, particularly at the technical level, and that of project management.

The period from 2008 to 2017 can be qualified as a phase of great changes, and of the highlighting of the challenges directly impacting the region's development.

During this period, four years were rainy (2010 and 2011 exceptionally so) and the others were dry. This situation led to new and exceptional phenomena in the reserve, notably forest fires of great severity and frequency. Between 2011 and 2012, the average number of fires grew from 14 to over 100 per year. The arrival of herds of camels and sheep (in their millions) from neighboring regions also grew massive. These herds have greatly disturbed systems of production through excessive use.

Thus, the fluctuation and succession of dry and heavily rainy years can favor or disfavor ecosystem conservation efforts, without forgetting their impact on the natural readjustment of ecosystem equilibria. Annual programs have been put in place by several actors (in particular the Agency for Water and Forests) to conserve ecosystems, soils and biodiversity.

The lack of rain negatively influences modern agriculture's management systems for use of underground water sources for irrigation, as well as coastal tourism and the habitual uses of urban centers. The adoption of aquifer contracts by establishments in this area, with the support of regional politicians, has been one adaptation taken to reduce the indirect effects of droughts, to insist on a more economical use of water for irrigation, and to plan and finance a system for reusing the water used in the large town of Agadir.

8.4. Changes and scalable trends in the ABR: from project territories to a territorial project

Possibly the most significant change with regard to short- and medium-term impacts on the dynamics of the ABR is the new constitution adopted in 2011. It gives an official place to civil society and to participatory actors, as well as propositional power with regard to policy choices around territorial governance, defense of equality and social solidarity. It holds an important role in the management of the environment and engages in the preservation of world heritage, in alignment with international standards.

The advanced regionalization project also has the effect of deconcentrating management powers and decision-making at the territorial level.

An important element in the evolution of the ABR is the creation of the National Agency for the Development of Oasis and Argan Zones (ANDZOA) in 2010. Its goals complement those of the South-West Regional Directorate for Water and Forests (DREF-SO) with regard to the application of three of the ABR's objectives. Within the ABR, the ANZOA concerns itself with the socio-economic development of the most disadvantaged areas, the problem of opening up the region and of supporting the development of the argan oil industry, while supporting efforts to preserve and rehabilitate the reserve's ecosystems, as well as encouraging scientific research.

In 2014, the Arganeraie was recognized as "Intangible Cultural Heritage of Humanity" by UNESCO. This recognition consolidated the Arganeraie's designation as a biosphere reserve.

In 2018, the Society for Regional Development and Tourism (SDRT) was created. Its primary mission is the stimulation of the ABR's hinterlands through the implementation of structuring projects, propelling the sustainable development of tourism, to the area's benefit.

Through this decade, the economic stimulation of the ABR has greatly changed at the local level through the development of the argan oil industry, boosted by state support and access to international markets, and through the development of a local product industry, particularly strong in aromatic herbs and honey. However, this evolution has been directly impacted by a territorial dynamic considerably affected by the movement of the active population from rural and mountainous areas towards towns in the plains and coastal centers that offer more lucrative employment opportunities.

Changes to the climate have been remarkable through this period, notably with the sharp drop in snowfall on the reservoirs of the ABR's summits, leading to clear decreases in flow in the rivers which feed the dams, all of this negatively impacting the agriculture reliant on the dams for irrigation.

With regard to the forested areas (80% of which are made up by argan trees), forest fires have become a serious problem. They are very likely the consequence of social factors, notably the rural exodus and the changes to systems of production. A significant increase in wild boar numbers has also been recorded in natural spaces, marked by movement between the forests and cultivated land. The remarkable return of the Cuvier gazelle has also been noted in the Anti-Atlas mountains.

Finally, the intensified movement of mobile herds towards the ABR's territories, more attractive in rainy seasons and years than in dry, has produced several tensions and conflicts at the local level regarding usage. In the past, rights holders within the Arganeraie periodically welcomed neighboring livestock rearers and shared, according to custom, the available resources. Today, however, the rise in population, climatic imbalances, social transformations and changes to systems of production have lifted such practices outside the realm of custom, and competitions over the use of resources have begun.

The total population of the provinces covered by the ABR has increased, between the two periodical reviews, from 2.78 million to more than 3.12 million inhabitants, with an average increase of 11%. It is important to note that at the level of local administrations, these socio-economic indicators show great disparities when compared to the regional or national averages. The equalizing of these areas thus remains an objective to revisit.

Following the creation of ANDZOA, the creation of the Trade Association for the Argan Industry (l'association de l'interprofession de la filière de l'argan) has been one of the defining events of this decade. This association gathers the entire industry together, from producers to exporters. The Federation of Rights Holders of the Arganeraie, a member of the trade association, benefits from a privileged position therein. It facilitates and encourages compromise and the development of partnerships, contact and dialogue between actors.

The Department of Water and Forests, represented by its regional directorates, is the main actor in the implementation of projects aimed at conserving and ensuring the sustainable development of ecosystems and their biodiversity. Fairly consistent programs have been realized through this decade, particularly with regard to the regeneration of the ecosystems surrounding the argan tree, the fight against erosion, and the tracking and preservation of the biodiversity offered by the natural habitats of the ABR.

The Regional Directorate of Water and Forests and the Fight Against Desertification (DREFLCD), as the ABR's managing body, ensures coordination and promotes the integration of projects at the planning level, and negotiations around certain kinds of intervention. The participative body, made up of all of the key actors, is the operational motor for this integration. Local associations and socio-professionals connected to the resource are frequently present, and their role in the participative body is highly important, even influential.

The management body, supported by the participative actors, oversees the development of territorial animation and stimulation and encourages advocacy in favor of the concept of the "biosphere reserve". A federation is structured around the ABR as a territorial project.

8.5. The ABR, complexities and improved governance

The analysis of trends in the ABR demonstrates the following notions:

– Climatic hazards and disturbances have become a determining factor in the efficient accomplishment of the conservational function.

– The scale of territorial contrasts necessitates a revision of the spatiotemporal scales of intervention and planning. The degree to which conservation and development functions can be reconciled is determined by the choice of a determined scale and a group of compromises between interests mostly concerned with development. To reconcile these two functions, it is necessary to reconcile territorial scales. At the territorial scale of close community management, the village is the base unit of local development (water, electricity, education). The base unit for the territorial scale of administrative management is the "commune", the governance of which is the responsibility of an elected council and an administrative structure following an action plan, with institutional support at the provincial level. The ABR is a textbook example, with eight provinces and more than 160 territorial communes.

– In conditions existing within a dynamic context influenced by national and international trends, the effective demand for development is greater than the demand for conservation.

– The evaluation of conservation and development efforts is often carried out in a linear and quantitative manner. However, the concept of the biosphere reserve was based on the concretization and reconciliation of conservation and development efforts. There are no scientific tools or practical approaches designed to measure the degree of reconciliation.

– The federation of actors in the ABR involves a shared collective interest in the values and benefits of this manner of treating and conceiving of the land. Similarly, this sharing involves the right and the responsibility to agree on one definition: concept, label or territorial project.

The ABR gives us an image of complexity, at two intersecting levels where the multiplicity of interest groups and the great variability of local specificities condition the fluidity and efficiency of its territorial governance within the objective of sustainable territorial development.

8.6. References

Boisvert, V. and Caron, A. (2010). La conservation de la biodiversité : un nouvel argument de différenciation des produits et de leur territoire d'origine. *Géographie, économie, société*, 3(12), 307–328 [Online]. Available at: https://www.cairn.info/revue-geographie-economie-societe-2010-3-page-307.htm.

Boujrouf, S. (2014). Ressources patrimoniales et développement des territoires touristiques dans le Haut Atlas et les régions sud du Maroc. *Revue de géographie alpine*, 102 [Online]. Available at: http://journals.openedition.org/rga/2259.

Duval, M. and Smith Benjamin, W. (2014). Inscription au patrimoine mondial et dynamiques touristiques : le massif de l'uKhahlamba-Drakensberg (Afrique du Sud). *Annales de géographie*, 697, 912–934 [Online]. Available at: https://www.cairn.info/revue-annales-de-geographie-2014-3-page-912.htm.

El Fasskaoui, B. (2009). Fonctions, défis et enjeux de la gestion et du développement durables dans la Réserve de Biosphère de l'Arganeraie (Maroc). *Études caribéennes*, 12 [Online]. Available at: http://journals.openedition.org/etudescaribeennes/3711.

Faouzi, H. (2017). L'arganeraie marocaine, un système traditionnel face aux mutations récentes : le cas du territoire des Haha, Haut Atlas occidental. *Norois*, 242, 57–81. doi: 10.4000/norois.6048.

François, H., Hirczak, M., Senil, N. (2013). De la ressource à la trajectoire : quelles stratégies de développement territorial ? *Géographie, économie, société*, 15, 267–284 [Online]. Available at: https://www.cairn.info/revue-geographie-economie-societe-2013-3-page-267.htm.

Gravari-Barbas, M., Bourdeau, L., Robinson, M. (2012). *Tourisme et patrimoine mondial*. Presses de l'Université de Laval, Laval.

Lacombe, N. (2015). Gentrification des activités dans l'Arganeraie : "l'éleveur" et la "concasseuse" comme figures sociales de déconstruction des agricultures familiales. *Bulletin de l'association de géographes français*, 92. doi: 10.4000/bagf.726.

Martin, J.Y. (ed.) (2002). *Développement durable ? Doctrines, pratiques, évaluations*. IRD Éditions, Paris.

OECD (1998). *Vers un développement durable : indicateurs d'environnement*. OECD Publishing, Berlin.

Tommasi, G., Richard, F., Saumon, G. (2017). Introduction. Le capital environnemental pour penser les dynamiques socio-environnementales des espaces emblématiques. *Norois*, 243, 7–15 [Online]. Available at: https://www.cairn.info/revue-norois-2017-2-page-7.htm.

UNESCO (2017). *Une Nouvelle feuille de route pour le Programme sur l'homme et la biosphère (MAB) et son Réseau mondial de réserves de biosphère*. UNESCO Publishing, Paris.

Wells, M. and Bradon, K. (1992). *People and Parks: Linking Protected Area Management with Local Communities*. World Bank, Washington, D.C.

9

Patrimonialization and Challenges to Sustainable Development within the Arganeraie Biosphere Reserve

9.1. Introduction

Biosphere reserves are ecosystems internationally recognized[1] by the Man and the Biosphere program. Their major task is to protect ecological spaces and their wealth of animal and plant life, as well as the biodiversity from anthropogenic pressures. In addition to this preservation process, they are concerned with safeguarding the sociocultural aspects of their territories, and promoting them so as to improve the living conditions of the communities therein, with an eye to sustainable development.

To this end, the Arganeraie Biosphere Reserve (ABR), situated in the center-west of Morocco, was designed as a tool for the protection and promotion of the resources it contains. It has implemented different development and heritage promotion strategies in an attempt to fulfill its objective of reconciling the economic, the ecological and the social, while respecting the interests of all of the stakeholders involved in the reserve.

Chapter written by Wahiba MOUBCHIR and Saïd BOUJROUF.

1 The "biosphere reserve" concept dates back to 1968 before the launch of the international and interdisciplinary MAB research program by UNESCO. One of the recommendations made during the UNESCO Biosphere Conference was to create an international and interdisciplinary research program, taking into account the problems of developing countries, which gave rise to the Man and the Biosphere (MAB) program in 1971.

However, heritage initiatives, witnesses to the logic of patrimonialization unfolding within the reserve, have seen an increase in the number of programs designed to manage natural and cultural resources, leading to a situation of environmental and societal stress. The propagation of actors (the state, civil society, the private sector, international cooperation), stakeholders in the same valorization process but with differing interests, has led to the excessive use of heritage resources. This has given rise to what some researchers (Michon et al. 2016) have called "overpatrimonialization", a kind of vulgarization of heritage due to an excessive exploitation, where the heritage property is distanced from its cultural value and acquires market connotations.

It is through these controversial situations that this chapter seeks to retrace the manner in which the process and the procedures of patrimonialization operate based on the apparatuses in place, and to interrogate the interactions between patrimonialization and the logic of conservation, questioning the mode of governance applied: to what degree have the actors involved in land management ensured the involvement of other users of the reserve's heritage resources in the decision-making process? Has the process of patrimonialization contributed to the tackling of the area's challenges and to the improvement of its population's living conditions, while abiding by the other imperatives around conserving biodiversity?

We will highlight the abuses of a patrimonialization emptied of its meaning through a lurch towards the market valuation of heritage resources. In the final part of this chapter, we will interrogate the possibility of putting into perspective a patrimonialization scenario in which new manners of governance could lead to the consideration of heritage as a sociocultural construction.

9.2. The ABR: a territory valued for the endemism of its heritage resources

9.2.1. *The ABR: the context of its creation and functions*

Created in 1998 with a view to safeguarding the ecosystems of Moroccan forests, and recognized by UNESCO in the same year, the ABR came to respond to the problem of degradation, a problem caused by increasing demographic pressure, encroaching urbanism, the threat of excessive grazing and the persistent threat of the loss of species and genetic diversity, particularly of endemic species. Indeed, one of the primary tasks of the reserve is to oversee the safeguarding of "an entire ecosystem or a representative biogeographical unit so as to defend its sustainability, or at least the continuity of the processes and mechanisms of the natural or semi-natural evolution of the ecosystem" (El Fasskaoui 2009). It is within this context that Morocco has defined a conservation policy for its threatened ecosystems.

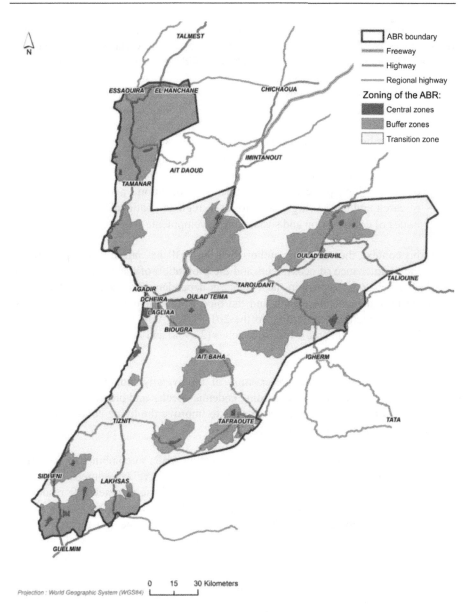

Figure 9.1. *The zoning of the ABR (source: South-West Regional Directorate of Water and Forests and the Fight against Desertification). For a color version of this figure, see www.iste.co.uk/romagny/biosphere2.zip*

The Arganeraie Biosphere Reserve encompasses 250 rural communes from the 10 provinces of Taroudant, Essaouira, Tiznit, Agadir-Ida Outanane, Inezgane-Aït Melloul, Chtouka Aït Baha, Safi, Chichaoua, Sidi Ifni and Guelmim. Its argan forests cover an area of approximately 830,000 hectares[2]. Its objectives come under the aegis of the Seville Strategy (UNESCO 1996), which defines the functions of each biosphere reserve:

> 1) contribute to the conservation of landscapes, ecosystems, species and genetic variation; 2) foster economic and human development which is socio-culturally and ecologically sustainable; 3) support for demonstration projects, environmental education and training, research and monitoring related to local, regional, national and global issues of conservation and sustainable development.

In this context, the managing authority of the ABR has put in place apparatuses aiding the maintenance of ecosystems and the promotion of the resources on which sustainable development within these areas is dependent. The zoning of the ABR has taken into account the wealth and the biodiversity of the units of which it is composed, and includes, according to the 2018–2027 ABR action plan, 24 central areas (2.2.% of the total area), 13 buffer zones (22.9%) and the remaining 74.9% are considered as transition areas.

This zoning concerns the preservation of biodiversity, notably within fragile areas where it is a matter of safeguarding endemic species and protecting them from human impact (central areas), and also aims to improve the living conditions of the local population in the two other areas.

Figure 9.3 presents the primary functions of the ABR: to combine the valuative discourse necessary for the qualification of the reserve's resources with the imperatives imposed by conservation through a rational use of renewable resources and a complete respect for sociocultural traditions. Following ANDZOA[3]: "it is not a matter of delimiting zones which would exclude, for the most part, human influence, but, on the contrary, of assimilating the latter's usufructuary claim in an integrational concept". The fulfillment of these functions led to the efforts, at various levels and on various scales, involving the local population and mobilizing other authorities: the Ministry of Agriculture and its decentralized bodies, Water and

---

2 ANDZOA website: www.andzoa.ma.

3 ANDZOA: the National Agency for the Development of Oasis and Argan Zones was created under the directives of the King of Morocco Mohammed VI in February 2010. It is placed under the aegis of the Ministry of Agriculture and Fisheries and works towards the elaboration and realization of integrated development projects for oasis and argan zones in collaboration with all local actors and in harmony with national and industry strategies.

Forests, the Ministry of Tourism, the Ministry of Culture, local authorities, and national and international NGOs. This was followed by several action programs launched by the state, or in partnership with international bodies: among others, the "Green Morocco Plan", the "Arganeraie Conservation and Development Project" and the "Moroccan National Biodiversity Strategy and Action Plan 2016–2020".

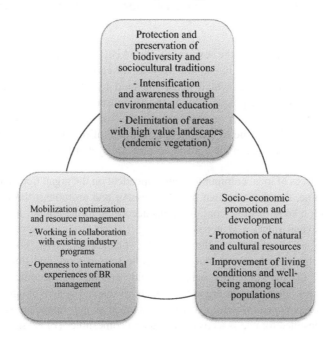

Figure 9.2. *The functions of the Arganeraie Biosphere Reserve*

This propagation of initiatives speaks to a real national will, and echoes international trends regarding vulnerable areas.

In this vein, let us recall that on the international scale, a cluster of actions[4] has suggested a particular interest in worrisome situations within fragile areas. This preoccupation has led to engagement in the form of partnerships (in terms of research, civil society, funding and project management techniques) between different bodies. It is a veritable godsend for southern nations, which certainly have much to learn from the leading countries in this area.

---

4 Among others: the Nagoya Protocol (APA), the Sustainable Development Goals (SDG 2030) and the Paris Agreement, from COP 21, for the protection of green spaces.

We also note that these exchanges resulted in the consideration of the argan tree by international organizations (FAO, GIZ, and the donors GCF, FA, FIDA, FEM, FFEM, the World Bank, etc.).

9.2.2. *The Arganeraie: a resource deposit under anthropogenic pressure*

It is certain that, given the extent of the ABR (approximately 2.5 million hectares)[5], the Arganeraie is home not only to its endemic species but also to a variety of agrarian resources which make the land to which they belong unique. However, emphasis will be placed only on the argan tree, the magical tree which has become the symbol of the reserve: especially since, from a purely economic point of view, the argan tree forms the true economic resource of the ABR (the extraction of argan oil and its importance in the agro-silvo-pastoral system) (Siminel et al. 2009).

Indeed, the lives of local communities are dependent on the argan tree, and since the beginning of time, they have known how to preserve and promote it through a series of rituals which take on diverse configurations according to the region. It is a heritage resource in its own right: it is considered "a gift from God and is represented by the population as the father of them all" (El Fasskaoui 2009). The unanimous recognition of this tree reveals its place in the collective memory and situates it at the origin of all of the practices which would go on to become subject to patrimonialization. Thanks to the argan tree, an entire adapted system of agriculture and apiculture was developed, which gave rise to coveted shared resources such as "the spurge honey destined for use as a cosmetic and as a comestible" (El Fasskaoui 2009).

Around the argan tree is woven a tapestry of knowledge and know-how which characterizes the process of its production. From the first stage of picking to the final stage of consumption, the argan fruit, in its raw or transformed state (through the production of argan oil, among other processes), gives rise to different uses, from which have originated ancestral practices handed down from one generation to the next. Between pulping, crushing, roasting over low heat, further crushing and mixing until the paste appears from which the oil is extracted, a know-how is permanently constructed, around which ties between community members would be woven through long-anchored rituals and traditions.

From this point of view, the argan tree is the foundation of an entire agrarian civilization built around this tree. In view of its characteristics, it would go on to be recognized by UNESCO as world heritage, and become the symbol of the Arganeraie Biosphere Reserve.

---

5 See: www.rbarganeraie.ma.

For its virtues, the argan tree has acquired a particular value at the national level, which justifies the state's and other private bodies' preoccupation with its promotion and preservation. Since the creation of the biosphere reserve, it has been a matter, on the one hand, of safeguarding the tree, as an example of world heritage, and the ecosystem to which it belongs, and on the other hand, of creating a dynamic of local economic development around its production. The ambivalent position of the state, particularly with regard to the management of the biosphere, emanates first and foremost from a wish to protect the argan tree, as well as from a response to the usufructuary claims of the local populations (ANDZOA).

The discourse[6] around preservation, which was at the very origin of the creation of the ABR and Morocco's other reserves, should be thought of as an imminent response to the different forms of human influence on these vulnerable spaces.

Under the influence of massive urbanization around Agadir and the demands of economic growth (modern agriculture, industry, mass tourism, the opening of new roads, etc.), several spaces have been cleared, to the detriment of the beautiful landscapes created by the argan-covered massifs and planes. Several authors have highlighted the impacts of this anarchic and wild urbanization on argan areas, a form of overexploitation of previously agrarian spaces which have been sold to real estate investors who care little for the matter of conserving ecosystems on the edges of the urban world.

For its part, the tourism industry has only exacerbated the situation by contributing to the development of a rural tourism which is poorly or hardly integrated. Besides, "for a long time the Arganeraie has felt the pressure of herds and of deforestation for energy or construction purposes", and the pressure of the rise in "poor practices", such as "those of shaking trees to gather fruit more rapidly [...] which limits regeneration opportunities" (Michon et al. 2016). In addition to these factors, we can note the development of agrobusinesses, and their impact on soils and water consumption to the detriment of other food crops.

It is to aid this increasingly alarming state of affairs that several initiatives have come to light for the containment of the risks caused by the overexploitation of the ABR's land. We can cite here the efforts of ANDZOA, of Water and Forests, and of other bodies to restore the argan forests from the verge of desertification, an initiative which contributed to the recovery of 101,487 ha in 2017. From 200 to 2007, 1,294 ha have been encompassed by the "Argan Tree" project, and 200,000 ha by the Green Morocco plan (Jadaoui 2012).

---

6 We allude here to the first intergovernmental scientific conference dedicated to the rational use of biosphere resources, which eventually gave rise to the MAB program.

It is also in response to other disagreements, caused by the strong media presence of the argan tree and its nutritional and cosmetic value combined "with threats of argan cultivation outwith Morocco or the usurpation of its name by foreign bodies" (Michon et al. 2016), that the state undertook a series of actions to promote and protect the argan tree and its territory by conducting a process of qualification and enhancement[7].

9.3. The ABR patrimonialization process

The idea of developing the ABR was birthed in response to several demands imposed by the fact of a handful of areas and their resources becoming the keystone for the entire region, and constituting a reservoir of argan oil production for national and international markets.

This examination of the qualification of areas through the promotion of their resources can be added to "the reflection led in 2007 into the updating of agricultural policies and, at the state level, an interest in rural heritage" (Michon et al. 2016), with an attached objective of "development in solidarity with small-scale agriculture" (CDGA 2009).

Following the same authors, rural heritage, in its different forms and contents, was not directly targeted by these early organizations, but "it appears as the meeting point between objectives and strategy in pillar II of the Green Morocco plan, in the form of concepts of locality, and of local products".

In the same context, the patrimonialization process as an apparatus is considered to have marked the beginning of the state's acknowledgment of the protection of properties and practices in which there is a collective interest, and of the necessity of maintaining its engagement with the local populations in terms of the management of the resources in question. Heritage has since been considered in the context of a territorial approach in which patrimonialization can be considered as a logic that favors the reinsertion of heritage properties into the new dynamic of sustainable development.

This new positioning of the heritage product places argan oil at the crossroads of two visible discourses: the discourse of the local product with symbolic attributes worth safeguarding, and that of the product highly sought after for its market value. The wish to reconcile these understandings of the heritage object has placed the state

---

7 We cite here ANDZOA's Territorial Development Strategy, outlined in 10 strategic documents and in 45 development programs. They amount to a global investment of 93 million dirhams (which is approximately 11.5 million per year over eight years) and hold ambitious goals for 2020.

under different constraints relating to the highlighting of procedures associated with the patrimonialization process.

Initially, efforts were focused on finding resources within the area with specific qualities. Argan had communally been raised to the position of flagship product – one which obscured others – around which the process of enhancement and protection had been constructed, based on the selection of values justifying its position as a heritage resource (historical, cultural, social, cosmetic, gastronomic, artisanal, etc.). This was followed by the identification of the practices (know-how) which had perpetuated its use within the community through the transfer of ancestral knowledge.

Furthermore, the reflection on the patrimonialization process within the ABR requires us to read from two constants: the approaches of the actors and the territorial approach.

The first step informs us on the nature of the different interactions between the actors involved in the enhancement process, and pushes us towards a reading of the logics which underlie their interventions. The patrimonialization process was initiated by Moroccan authorities with foreign cooperation, with the aim of promoting and above all protecting the argan tree and its territory from the threats that its foreign cultivation may provoke. To this first concern was added the necessity of associating the resource's production process with a brand image, which would allow it to assert itself as a local product with high added value: patrimonialization demands the "diverse rights of intellectual property [...] labels, certifications, provisions relating to the recognition of indigenousness or of the right to culture" (Boivert 2013).

A geographical indication system was thus envisaged for the official identification of argan oil. For this reason, the Moroccan Association for the Geographical Identification of Argan Oil (AMIGHA) was created in January 2008, with the goal of encouraging the extension of the national market for the oil through a mark of quality, a Protected Geographical Indication (PGI)[8]. It guarantees product quality by certifying the trajectory of and the standardization of the stages characterizing its production[9].

The second step consisted of the implementation of a dynamic of actors around the local resource through the creation of new structures (cooperatives, economic interest groups), the involvement of private companies and decentralized

---

8 The PGI was obtained in 2010 by ministerial decree, promulgated by Dahir no. 1-08-56, of May 23, 2008, following the publication of law no. 25-06, relating to distinctive signs of origin and quality.

9 See: www.rbarganeraie.ma.

international cooperation through NGOs: "the very first argan cooperatives initiated by the GTC within the framework of the PCDA"[10] (Romagny et al. 2016).

We would suggest that the preparation of this stage of the ABR's heritage enhancement, organized exclusively by the state, looks like a consensus between the different stakeholders. Patrimonialization is thus reduced to the construction of the argan industry and the opening of a promotional market, the profits of which are not seen by most of the local population. The cooperatives, which should have been placed at the center of production to guarantee the sustainable management of the heritage resource, were instead given subordinate roles:

> the Arganeraie [...] uses underpaid local women to collect and crush the argan fruits and [...] has moved a large part of its transformation and commercialization operations outwith the area, and moves the profits outwith the Arganeraie (Romagny 2010).

These practices allow us to suggest that the state has found itself unable to reconcile the enhancement of heritage and its marketization. It has become evident that these logics cannot coexist. Patrimonialization cannot work when the intentions behind it are purely commercial and work to strip the argan tree of its symbolic charge by gradually altering the mechanisms of its production. With the proliferation of foreign investors, the discourse around the collective management of heritage resources has become illusory.

Patrimonialization through this play of actors has oscillated between visibility and invisibility, beginning as an articulate state discourse, oriented towards a logic of enhancement and conservation and with the implementation of apparatuses adequate for the management of heritage. Our reading of the play of actors involved in the heritage cause has shown the presence of rather hierarchical relationships between the different intervening groups with their divergent interests, and a quasi-absence of dialogue between them.

Instead, we are seeing power relations in which small structures (cooperatives, rights holders, locals) have found themselves divested of their property, and in which large private companies and their suppliers, wholesalers and manufacturers monopolize most of the profit. To this end, some authors speak of "the banalization of heritage". When heritage finds itself "tied to market valuation [...] we end up in situations where the construction of the image overflows from the processes of collective construction and leads to an authentic illusion" (Michon et al. 2016).

---

10 PCDA: Arganeraie Conservation and Development Project.

Similarly, heritage has been placed at the center of the ABR's territorial development strategies through a reliance on enhancement activities such as tourism, in such a way that it directly profits the local population. Nobody can deny the dialectical relationship between tourism and heritage, nor their co-constitutive character, which has largely served as a base for area development projects. According to this logic, the ABR bet on this sector to enhance the reserve's resources, while trying to reconcile touristic interests and the constraints imposed by the conservation of biodiversity. This is a challenge for the area, which has succeeded in attracting significant tourist flows, thanks to a highly diversified appeal essentially based on the notoriety of the ABR and its resources: local products, the diversity of its programs, based on the discovery of its varied landscapes and its cultural riches. Its important infrastructures (airports in Agadir and Essaouira; accommodation and leisure infrastructure) have greatly contributed to the attractiveness and competitiveness of the ABR's territories.

However, despite the profit the ABR and its territories were able to pull from tourism, new regards would go on to alter the relationship between locals and the heritage previously seen as a basis for investment. These transformations in the way heritage properties are understood would unfortunately lead to some highly debatable practices: certain investors, under the pretext of wishing to renovate an old building seen as heritage, ended up disfiguring it, misleading tourists seeking authenticity. Similarly, tourists transformed into investors: certain authors have spoken, in this context, of the relationship between "patrimonialization and otherness" while making the link between the manner in which the regard of a foreigner can transform the heritage object: Yerasimos (2006) speaks of "imported patrimonialization", attributing to it new values when it had fallen into disuse. Meanwhile, this view of otherness can, in the absence of legislation and supervision, veer towards a "vulgarization of heritage property".

With regard to governance, the ABR has seen an abundance of actors whose diversity has generated conflicts of interest, revealed a lack of coordination (Aboutayeb 2014) and a lack of transversality in the relations between the different participants, and spoken of "overlaps of responsibilities and prerogatives". These factors represent "a risk that the network of actors may collapse because of insufficient conflict management mechanisms, and a lack of any universally recognized organizational power". The question is whether or not an agency, like ANDZOA, has the tools and profile necessary to install a system of governance that could support tourism within the ABR.

It turns out then that the challenge to the areas which have chosen patrimonialization as the process likeliest to tie the enhancement of the area to its protection is immense. Should we be seeking a sustainable territorial development that does not involve a life cycle which moves from specification to banalization, or from the specific to the generic? How can these processes be relaunched, so as to focus on the specific object as regenerated within an inclusive and convergent territorial project? To answer these questions, the following section will address paths of governance that can make the ABR a space of dialogue and mediation in favor of sustainable territorial development.

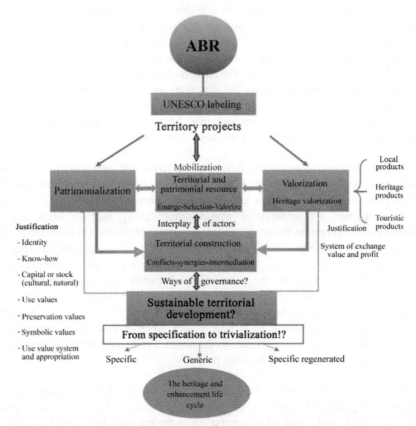

Figure 9.3. *Life cycle of patrimonialization and enhancement in the Arganeraie Biosphere Reserve*

9.4. Paths of governance for the integrated management of the ABR

By "paths of governance" we mean possible management methods that could be offered to resource managers, allowing them to ensure coordination between the different actors in the relevant areas.

Our proposition is based on the biosphere reserve governance model from Beuret (2011). The value of the study in question is that it allows, based on the exploration of a variety of management methods, for the establishment of a governance model based on innovation and dialogue, and the promotion of a visibility at the area level which is relative to the choice of a given mode of governance. We do not claim to adapt every path of governance analyzed within the framework of this study, but have rather chosen to focus on those based around dialogue, which are able, it seems to us, to be reinscribed into the subject of our study (patrimonialization), and which could thus have significant repercussions for our area.

Through our summary analysis of the patrimonialization process within the ABR, we have hoped to show the difficulties that territories have in reconciling the preservation of the environment and the practices stemming from territorial resources. We have broadly noted that, despite existing apparatuses, it has been difficult to unite actors around a project of shared interest. The experience of patrimonialization has only reinforced the most powerful actors, and weakened poorer actors and the rights holders, above all the women represented through their cooperatives.

The major challenge for an area is to reconcile actors with different objectives, which "supposes the breaking-down of the barriers existing between them" (Beuret 2011). In other words, it is a question of amplifying paths of dialogue so as to reduce tensions and stressful situations that may be provoked by the appearance of an actor with intentions which seem difficult to understand, or which are mysterious. The governance of protected areas is "a process not only of coordinating actors, but of appropriating resources and constructing territoriality" (Leloup et al. 2005). "It offers itself within territories" (Beuret 2011) as "the process of structuring the compatibility of different methods of coordinating actors in geographical proximity" (Gilly et al. 2004).

Furthermore, the Seville strategy (UNESCO 1996) proposes to "[s]urvey the interests of the various stakeholders and fully involve them in planning and decision-making". Thus, the challenge of governing a biosphere reserve is to deal with the different contexts related to the distribution of its zones of action (central area(s), buffer zones and transition areas). Such technical zoning always poses the problem of acceptance and "understanding" on the part of the local population. But it is always a question of adapting modes of governance to nature and to the contingencies of the area in which we plan on establishing them, hence the variety of paths of governance, specifically contracts, incentivization, deliberation and institutional rearrangements.

9.4.1. *The path of the contract*

This is an apparatus considered as a type of "contractual coordination". "It combines a trading mechanism (a service rendered in exchange for payment) with a consultation between two parties". An area's managing authority authorizes under contract access to use a resource and imposes in exchange "a contribution" to the conservation of the environment, with an offer of compensation "following the rules fixed by the managing authority of the reserve" (Beuret 2011). Users thus become the environment's protectors, and are involved in functions which make them aware of and responsible for the fragility of the area. A second condition promotes access for service providers who engage in production activities (as in the case of private companies in the Arganeraie reserve) while contributing to conservation.

Access to resources is in principle regulated by the reserve's managing bodies (the Administration of Water and Forests within the framework of the ABR, with the mediation of ANDZOA, habitually takes on this role). This kind of contractual coordination can be supplemented by accompanying actions from the managing authority around concerted projects (as in the case of touristic or heritage promotion). The value of this path of governance is found in "the contributive value" of actants (particularly locals) who become the allies of the regional or central power, which can diminish recourse to "illegal activities", notably in exclusionary zones.

9.4.2. *The path of deliberation: consultation and concertation*

Governance is by definition conceived of as "a deliberative act", a review considered in consultation with other people. It is based on "mechanisms of information and consultation" highlighted throughout the implementation of a project. However, according to the same author, different scales of reference appear as "informative, contributive, and interactive consultations", the quality of which rests on the amount of time accorded to participants to adapt to the project and to progress towards a collective learning. This returns us to the patrimonialization project within the ABR, in which consultation took place on an ad hoc basis, without the parties involved (particularly locals) having time to adapt to the project. Consultation, in our case, was opened upstream, reducing it to advisory consultation, ideas from other participants not being taken into account during the construction or implementation stages of the project.

The path of deliberation demands a dialogue prior to any project, taking into account the ideas expressed by all consulted parties, which should be discussed in order to internalize the information provided.

Concertation is defined as the act of sharing common objectives, which allows "actors to respond to a gamut of problems that they will encounter together or in their collective activities, based on common references". In the construction of projects, concertation is a "participative research" procedure that should immediately engage local communities in the gathering of information regarding know-how and heritage properties. It is a matter of awakening them to their values and initiating them into an appropriate use of their nature. It is also a matter of preparing them for a fairly communitarian management based on the creation of village committees supported by "a participative research team, which acts as a facilitator".

9.4.3. The path of incentivization

The path of incentivization is based on the encouragement of actors operating in the same territory and who involve themselves in sustainable environmental development activities. This governing apparatus is initiated by management authorities who are decentralized, opening a space for actors engaged in different projects to invest in conservation activities within the reserve. These projects are not market-oriented and aim for the involvement of everyone in the environmental, socio-economic and cultural causes.

The role of the central power is to promote conditions of accompaniment, putting necessary expertise at the disposal of interested actors (information, technical management models, awareness), promoting a space for dialogue that is "active and innovative with regard to nature conservation".

9.4.4. The path of institutional rearrangement

This path consists of the creation of bodies given the objective of facilitating collective development and conservation activities within the biosphere reserve. This is the situation of ANDZOA within the ABR. Management authorities can also cooperate with other public bodies when they are involved in collective actions.

In this respect, we can cite the example of actors involved in tourism, such as the CRT (Agadir Regional Tourism Council), the CRSMD (Souss Massa Draa Regional Council), CPTs (Provincial Tourism Councils) and the RDTR (Rural Tourism Development Network). These bodies, whose presence in the same territory has thus far produced tension and latent conflict, are supposed to interact within a governance of institutional rearrangement benefiting from conditions conducive to participative action. To this end, Beuret speaks of "the redistribution of certain prerogatives and/or financial resources" by the reserve's central authority so as to stimulate sustainable development. In this respect, ANDZOA has adequate potential to play this "facilitative" role in its turn.

Finally, regardless of which path of governance is chosen, it is important to know how it will be implemented. This leads us to the choice of actors tied to the governance. Beuret speaks of the "process of legitimation" of certain actors based on two conditions: the first is linked to the use of resources and the activities stemming from this use, and the second recognized inhabitants of the area as the rights holders or legitimate representatives of the area. This position confers on them "the right to expression and to property". In these two cases, it is a question of "two very different ways of distributing power, of reinforcing certain territorialities or weakening others, of selecting social networks for the governance to enforce, creating very different socio-organizational configurations" (Raffestin 1980).

We see here that the paths of governance are very different, a diversity which, in the case of the ABR, has been insufficiently exploited, despite putting in place of deliberative apparatuses and despite the selection of legitimate actors in the first place. Centrality is the only present configuration, and other bodies are only invited through this centrality, through highly ad hoc activities. The multidisciplinary management that follows the Arganeraie's major axes of development and conservation seems to have fragmented the area into spaces of competition between different actors whose cooperation with regard to action is currently difficult to engage.

9.5. Conclusion

At the end of our reflection on patrimonialization within the ABR, it appears that multiple apparatuses and strategies appear to promote a true enhancement of heritage: namely, geographical indications, for market promotion, and incentives for different types of engagement (cooperatives, NGOs, locals, private companies, etc.), around the heritage object.

In every case, the massive interest in the exploitation of local products ended up stripping the heritage object of its values. This stems from the appearance of private investors who succeeded in inserting the fact of heritage (a system of values and ancestral cultures) into a model of capitalist exploitation. This conception deals with the objectives of the state, which was more preoccupied with "a restructuring of the traditional sector both upstream and downstream" (Michon et al. 2016), thus revealing purely structural intentions. Thus, we can suggest that the arrival of territorial projects did not lead to the establishment of any real governance, especially if we consider the coordination mechanisms deployed to this end, notably with regard to the formation of decisions. The state contented itself with incentivizing activities through the engagement of actors in the area, or even the creation of other coordination structures to stimulate their participation. Its omnipresence at the territorial level has always placed it as the direct manager of the reserve, allowing it to influence other actors by mobilizing different apparatuses (training, awareness, accompaniment).

Based on these observations, the challenge for the ABR should be to explore other paths of governance, exploiting their complementarity while reconsidering the position that actors engaged to this end should occupy. Local communities, being at the center of all reflections linked to development, should see their legitimacy reinforced by the mode of governance. The structures of local actions should be constructed around the interests of users, and integrated into true dialogical bodies along with other actors in the area.

The incentivization of contribution to the patrimonialization process should not be limited to a few assignments in the heritage object's chain of production. The challenge is, on the one hand, to reinvent the participative approach based on the traditional fabric (of socio-territorial unities: tribes, factions, etc.) so as to inclusively organize users within projects of a shared interest, and, on the other hand, to persuade these users to reconstruct their ties to the past through individual and collective leaning, allowing them to immerse themselves in a culture they will thus be able to reproduce and transmit.

9.6. References

Aboutayeb, H. (2014). La Réserve de biosphère de l'Arganeraie : un nouvel écoterritoire touristique au sud du Maroc. *Pasos revista de turismo y patrimonio culturel*, 12(4), 915–922. doi: 10.25145/j.pasos.2014.12.066.

Beuret, J.E. (2011). La mise en dialogue de la nature dans les réserves de biosphère : les voies de la gouvernance. *Bulletin de l'association de géographes français*, 88(4), 459–470.

Boivert, V. (2013). Marchandisation ou Patrimonialisation ? L'économie de la biodiversité en perspective. In *Effervescence patrimoniale au Sud. Entre nature et société ?*, Juhé-Beaulaton, D. (ed.). IRD Éditions, Marseille.

El Fasskaoui, B. (2009). Fonctions, défis et enjeux de la gestion et du développement durables dans la Réserve de Biosphère de l'Arganeraie (Maroc). *Études caribéennes*, 12 [Online]. Available at: http://journals.openedition.org/etudescaribeennes/3711 [Accessed 14 December 2020].

Gilly, J.P., Leroux, I., Wallet, F. (2004). Gouvernance et proximité. In *Économie des proximités*, Pecqueur, B. and Zimmerman, J.B. (eds). Hermes Lavoisier, Paris.

Jadaoui, M. (2012). Savoir-faire local et développement durable : pistes de réflexion à travers l'exemple de l'Arganeraie. PhD Thesis, Université Ibn Zohr, Agadir.

Leloup, F., Moyart, L., Pecqueur, B. (2005). La gouvernance territoriale comme nouveau mode de coordination territoriale ? *Géographie, économie, société*, 7, 321–332.

Michon, G., Berrian, M., Romagny, B., Skounti, A. (2016). Les enjeux de la patrimonialisation dans les terroirs du Maroc. Les territoires au Sud, vers un nouveau modèle ? Une expérience marocaine. Faculté des lettres et Sciences Humaines, Université Mohammed V de Rabat, Rabat.

Raffestin, C. (1980). *Pour une géographie du pouvoir*. LITEC, Paris.

Romagny, B. (2010). L'IGP Argan, entre patrimonialisation et marchandisation des ressources. *Maghreb-Machrek*, 202, 85–114.

Romagny, B., Boujrouf, S., AitErrays, N., Benkhallouk, M. (2016). La filière huile d'argan au Maroc. Construction, enjeux et perspectives. Report, Faculté des lettres et Sciences Humaines, Université Mohammed V de Rabat, Rabat.

Simnel, R., Michon, G., Auclair, L., Thomas, Y., Romany, B., Guyon, M. (2009). L'argan : l'huile qui cache la forêt domestique de valorisation du produit à la naturalisation de l'écosystème. *Autrepart*, 2, 51–73.

Yerasimos, S. (2006). *La deuxième mort du patrimoine. Villes réelles, villes projetées*. Maisonneuve et Larose, Paris.

Websites

Agence Nationale pour le développement des zones oasiennes et de l'Arganier (n/a). ANDZOA [Online]. Available at: http://andzoa.ma.

Centre d'échange d'information sur la biodiversité au Maroc (n/a). Stratégie et Plan d'Action National sur la Biodiversité [Online]. Available at: http://ma.chm-cbd.net/implementation/snb_ma.

Réserve de biosphère de l'Arganeraie (n/a). Zonage de la réserve de biosphère de l'Arganeraie [Online]. Available at: https://rbarganeraie.ma/zonage-de-la-reserve-de-biosphere-arganeraie.

Wikipedia (2022). Réserve de biosphère [Online]. Available at: http://fr.wikipedia.org.

10

The Oasis du Sud Marocain Biosphere Reserve: Challenges and Issues for the Durability of Water Resources

10.1. Introduction

For millennia, water has been managed within arid zones, in the search for a balance between the resources themselves and the needs of inhabitants. Ancient civilizations came to master this vital resource rationally enough, to exploit it, share it out and to use it in diverse and ingenious ways, always with an eye to ensuring the stability of the community. Today, water resources are increasingly scarce, and regulation of its use is increasingly complex.

In Morocco, oases appear as spaces with strong potential for the development of their inhabitants. Their position and extent at the edge of the desert make them privileged areas which, throughout history, have produced a mosaic of cultures and a melting-pot of populations, leaving behind unrivalled cultural and architectural heritage.

In the oasis, water resources are, however, incredibly fragile, dependent on fluctuations in rain levels over space and time. This vulnerability is made all the more tangible through the influence of climate change. Morocco has, however, implemented public policies and strategies to take oases into account. Since 1996, Morocco has been engaged in, and in 2001 welcomed, the Conference of the Parties (COP 7) of the United Nations Framework Convention on Climate Change in Marrakech.

Chapter written by Lahcen AZOUGARH and Ahmed MOUHYIDDINE.

The great territorial diversity of Morocco forces it to address a multitude of different challenges, such as preserving resources, improving the attractiveness of different areas and seeking to be competitive within the standards of a healthy space. The National Strategy for Sustainable Development (SNDD) (2015–2030) thus proposed a reinforcing of activities promoting the most sensitive spaces, including desert zones and oases. These spaces represent more than 40% of Morocco's land and are poor, threatened by salinization and silting, poorly equipped with regard to infrastructure, and rendered ecologically vulnerable by the harshness of the climate, the scarcity of water, the risk of degradation as exacerbated by overgrazing, etc.

This chapter is a reflection on the specificities of the Oasis du Sud Marocain Biosphere Reserve and the state of its water resources: we will attempt to examine the relationship between economic development and the protection of natural resources, particularly water, through Morocco's experience in preserving this biosphere reserve.

10.2. Specificities of the Oasis du Sud Marocain Biosphere Reserve and the question of water

Geographically, the Oasis du Sud Marocain Biosphere Reserve (OSMBR) includes the Drâa, Ziz, Gheris and Guir drainage basins, the main oases of the south face of the central and eastern High Atlas mountains, the entire south face of the central and eastern High Atlas mountains and the Ouarzazate, Tinghir and Errachidia basins, hollows which follow the great tectonic fault line to the south of the High Atlas mountains. It was designated a biosphere reserve by UNESCO thanks to the richness of its biodiversity, its cultural and historical value, and the presence of know-how that relates to the challenges of sustainable development.

The great diversity of its substrates, combined with its strong variation in altitude, its water gradients and its extreme temperatures, creates extremely varied ecological systems throughout the OSMBR. This environmental variability has given rise to an ecological diversity that makes the High Atlas area a center for diversity and for an endemism that is exceptional at the continental scale. Another factor contributing to the OSMBR's biological diversity is its position in the contact zone between the Mediterranean, Irano-Turanian and Saharo-Arabian floristic regions. Here, we can find circumpolar species such as *Parnassia palustris* alongside clearly African contributions like *Maerua crasifolia*; there are Mediterranean green oak forests and continental steppes covered in stipa and artemisia. The floristic richness of the OSMBR does not result solely from the superposition of different regional flora, but is characterized, rather, by a remarkable individuality. Endemism rates increase from the Saharan to the oro-Mediterranean zones, and more than 80% of the area's flora appear in this category. In terms of fauna, the area's inhabitants are also extremely varied, particularly thanks to its highly contrasted topography, and the OSMBR's territories thus offer strong representative value, for example, with regard to Moroccan vertebrates.

Moroccan oases also offer the best conditions for the sustainable development of natural resources. Indeed, between habitats built within ksars and irrigation systems like acequias and khettaras, an incredible range of systems and modes of growing are still in vogue.

The area's agriculture is distinguished by a strong diversity of farming systems based around polyculture, associated or not with livestock rearing. The practice of growing multiple crops on the same plot is very popular, and the choice depends neither on the size of the establishment, nor on the region, but on socio-economic considerations born from the permanent concern of farmers around securing and diversifying their revenue streams.

Terraces lend oases a typical, and rather attractive, agricultural beauty. The OSMBR, endowed with diverse landscapes and a hospitable population, thus offers a unique touristic product within Morocco. It benefits from an international tourism trend which is seeing visitors, particularly Europeans, expressing an increasing interest in desert regions and in the strongly humanized heritage of oases.

The biosphere reserve's border with the Sahara Desert makes aridity a structural phenomenon, leading to an almost constant desertification. This problem is amplified by the climate, characterized by a long and extremely hot summer period (May–October) associated with scarce and irregular rain. The rains that often appear at the end of autumn and the beginning of winter are insufficient for the regeneration of the area's vegetation, which grows mainly in spring. These rains are often torrential, increasing the likelihood of run-offs and the erosion of arable land. Aridity is thus a major handicap to the economic development and ecological integrity of the reserve. The situation is worsening, and no environmental or agricultural project can stop it without taking into account this climatic variability.

The pressure on the environment, notably on its water, is strong, threatening the area's ecosystemic functions; the question of the sustainability of its water resources affects biodiversity, and a great number of people. The sustainable management of water resources thus represents a key challenge for community development. The OSMBR is an interesting example of the role that biosphere reserves can play in the creation of water sustainability.

10.3. Regional development and the deterioration of water resources

The establishment of a biosphere reserve represents a real challenge: that of putting in place a planning system that can contribute both to the protection of the environment and to territorial development. This organizational dimension makes it unique: the specific and particular management of a shared potential, by a local population desiring a form of socio-economic development adapted to its needs. It puts local populations in a position of divergence from state programs and those of economic investors which are likely to affect the ecological and cultural values of the biosphere.

Indeed, the water available under the water resource plan of the OSMBR's basins does not cover the water needs of the palm groves, and irrigation uses excessive amounts of water from deep (Plio-Quaternary) aquifers. Following the Study of the Master Plan for the Management of the Water of the South-Atlantic Basins (PDAIRE 2011), water table levels have been decreasing since March 1993. Renewable reserves have also seen, in the last few years, a deficit of infiltrating rain caused by a 50% reduction in precipitation. The shrinking of this water layer has also caused a sharp weakening of the flow rates of springs. This is manifesting in an increased reliance on the aquifer to satisfy water needs.

For several decades, the will towards sustainable development has had to face up to modernism, individualism and the proliferation of consumer activities involving water under the pretext of territorial development. Particularly in agriculture, the uncontrolled proliferation of extremely deep boreholes has put into question ancient equilibria. Today, the number of active boreholes is 5,790, where 10 years ago there were only a few hundred. More and more, aquifers are being dried up, and the very existence of some oases are threatened, an example being the Taghbalt Oasis (in the south-west), where water levels in some areas have reached more than 280 m, especially following the exploitation of a borehole by the mining industry. The exploitation of such an aquifer can be economically justified, but it involves the destruction of an example of secular heritage for the sake of only limited profits. The challenge of keeping a particularly close watch on sensitive areas is thus a strong priority. In addition to agricultural and mining activities, this excessive pumping is also supplying tourist infrastructure developed within the desert's limits, further contributing to the reduction of aquifer levels in all of the oases of the biosphere reserve. Added to this is uncontrolled tapping by new agricultural enterprises appearing on the periphery of the oases, in the OSMBR's transition zone, in detriment to the mother palms.

10.4. Challenges and complexities of water resource management within the OSMBR

The management of water resources within the Oasis du Sud Marocain Biosphere Reserve is not an easy task, the variables which enter the game being highly complex. It raises a number of questions and challenges. The agricultural policy launched in 2008 (GMP: Green Morocco Plan) aims to encourage farmers through financial aid grants, reaching up to 100% in the case of private properties, without environmental impact studies for the projects. It is a centralized policy, and this encouragement thus covers entire portions of the reserve's land, a land which is nevertheless highly contrasted and varied. This centralization and standardization of policy pushes farmers to invest in profitable non-native crops, without concern for the necessary water resources. The changes in the ways water resources are used and accessed between actors foreshadow the appearance of social conflicts of interests (such as ethnic or tribal affiliation) and spatial ones (location upstream or downstream of the drainage basin).

Biosphere reserves must be managed and protected in a sustainable way by appropriate legislative, regulatory or institutional apparatuses, and in clear consensus with customary law (*Azerf* in Amazigh; *Orf* in Arabic). Customary law regulates many water management practices that modern law has not yet addressed, this panoply of texts being compiled by the inhabitants of oases themselves and not by the jurists of the centralized state. They take scrupulously into account the details of the area, considering the plot, the tribal faction, the lineage and the availability of the resource – and the principle of solidarity between upstream and downstream – to ensure the sustainability of the landscape and the environment.

The question of water resources cannot be reduced down to a question of biodiversity, nor to a question of economics; it is a factor of social stability, and indeed potentially one of social instability (Houdret 2008). A holistic understanding of water management is necessary to take into account interactions between the environmental, social, economic and political aspects that make up a biosphere reserve. It is essential to note that "water is an acknowledgement of social ties and a mark of non-exclusion" (Bougerra 2003); it represents a uniting force among the biosphere reserve's population. This unifying aspect facilitates the mobilization of actors, and makes engaging them in planning, proposals or decision-making a more efficient and flexible process. Failing this, water can mend feelings of exclusion and paper over present or latent social conflicts within the area.

The complexity of its management and the fragility of the area necessitate the creation of an institution to track and evaluate water resources: an oasis observatory could track trends and identify the commitments of different actors. This institution, with its programs, could be supported by specific mechanisms and funding tools and by a particularly proactive land management policy. Within this establishment, a research laboratory could support the scientific supervision of doctoral theses on the biosphere reserve focusing on territorial diagnostic, the integrated management of resources, their environmental evaluation, the management of conflicts and the elaboration of research problematics adequate to the needs of the OSMBR. All of these propositions would help to bring the daily running of the reserve closer to the function of the biosphere reserve, to understand the interactions between humans and the biosphere and between social and natural systems within the framework of the Man and the Biosphere (MAB) network.

Scientific research within a biosphere reserve can aid the analysis and management of conflicts, by supplying information destined to supplement the dialogue so as to achieve a collective construction of solutions (UNESCO 2007). This knowledge would bear primarily on the use of resources and of biodiversity (understandings of behaviors, of customary law, etc.), on the perception of actors, on local knowledge or on good practices. Finally, the work of raising awareness of the MAB program remains tepid; its popularization could change understandings of, or even change the hand that is dealt with in regards to the scarce, shared resource that is water. Common property is difficult to manage; it requires collective action, as the pursuit of personal interest often destroys projects created by the community.

10.5. Conclusion

It is essential that authorities, NGOs, scientists and users of the biosphere reserve come together to understand global trends and changes (both environmental and social) with an eye to the suitable implementation of projects adapted to the specificities of each sub-territory within the biosphere reserve. The specificity of approaches must be increased within a framework of multidisciplinary studies and with the participation of all concerned (the indigenous population, the scientific community, ministerial departments, technical and financial partners, NGOs, etc.), so as to share with them the vision constructed by the group, and to arouse their interest and involvement in a network-oriented logic. This will allow economic partners to be shown that their involvement constitutes a brand image tied to a specific, unique and labeled area. It would also be judicious to update data on each area within the reserve (every 10 years), and to define the new challenges thus presented, for the sake of an integrated and exhaustive evaluation of everything making up this space, thus updating and allowing for the elaboration of a dynamic diagnostic in the form of a periodic environmental audit.

From a practical point of view, academic expertise and cross-sectional research remain indispensable, as well as the involvement of all of the relevant actors (regardless of their importance or relationships) in every stage of environmental prospecting.

10.6. References

Blandin, P. (2004). Développement durable et patrimoine naturel : une histoire commune. In *XXIIe congrès de l'AMCSTI (Association des musées et centre pour le développement de la culture scientifique, technique et industrielle), Parc naturel régional du Luberon*, 16–18 June.

Bouguerra, M.L. (2003). *Les batailles de l'eau pour un bien commun de l'humanité*. Enjeux Planète, Paris.

Cibien, C. (2006). Les réserves de biosphère : des lieux de collaboration entre chercheurs et gestionnaires en faveur de la biodiversité. *Natures sciences sociétés*, 14, 84–90.

Houdret, A. (2008). Les conflits autour de l'eau au Maroc : origines sociopolitiques et écologiques et perspectives pour transformation des conflits. PhD Thesis, Université de Duisbourg et Esen, Duisburg.

UNESCO (2007). Le dialogue dans les réserves de biosphère : repères, pratiques et expériences. Technical notes, UNESCO, Paris.

PART 2

Issues and Case Studies in the Southern Mediterranean

Part 2

Introduction to Part 2

This part presents several case studies which show that, between the powerful aspirations towards development held by their populations, the impact of human activities and the need to conserve their biological heritage, biosphere reserves in the southern Mediterranean are faced with issues which are difficult to reconcile: agricultural pollution, forest fires, water usage, all in a context of climate change, and so on.

Several chapters present difficult situations in Morocco, notably within the Intercontinental Biosphere of the Mediterranean. They address both the social innovation represented by a social economy based around solidarity and the mediatization of biosphere reserves as tools for change. One chapter analyzes the importance of history for conservation policies in Algeria. Finally, in Libya, in its position as a framework for the training of architectural students, the Jabal Moussa Biosphere Reserve has allowed for their initiation into problems of conservation and sustainability, and the development of innovative projects, as pedagogical in nature as architectural.

Introduction written by Catherine CIBIEN.

11

Pesticide Residue in the Waters of the IBRM

11.1. Introduction

Pesticides are chemical compounds primarily used to protect crops. They are divided into different categories based on their use (insecticides, herbicides, fungicides) or based on their chemical structure (organochlorides, organophosphates, triazines, carbamates, triazoles, etc.).

The International Code of Conduct on the Distribution and Use of Pesticides (FAO 2002) defines the term "pesticide" as:

> any substance or mixture of substances intended for preventing, destroying or controlling any pest, including vectors of human or animal disease, unwanted species of plants or animals causing harm during or otherwise interfering with the production, processing, storage, transport or marketing of food, agricultural commodities, wood and wood products or animal feedstuffs, or substances which may be administered to animals for the control of insects, arachnids or other pests in or on their bodies.

Used to agricultural or nonagricultural ends, pesticides were widely used between 1960 and 1984, particularly in the domains of agriculture and of public health for the eradication of certain diseases. However, pesticides are the source of a

---

For a color version of all the figures in this chapter, see www.iste.co.uk/romagny/biosphere2.zip.

Chapter written by Hind EL BOUZAIDI, Fatimazahra HAFIANE, El Habib EL AZZOUZI and Mohammed FAEKHAOUI.

"great number of the ills of the modern world" (Pouradier 2010). In reality, the majority of these substances remain dangerous to human health, particularly organochlorides, and are as harmful to animals as to humans. These substances can affect the skin, the eyes, the liver, the kidneys, the cardiovascular system, as well as the reproductive system. However, following the discovery of their potentially toxic effects in 1972, several countries across the world have placed blanket bans on their use (Sapozhnikova et al. 2004).

Pesticides can contaminate both subterranean and surface water deposits. Different authors have thus studied the dispersion and transfer of pesticides into the environment through several routes, particularly through water.

This diffusion of pollutants is linked to the (above- or below-ground) transfer of pesticides by rains and by irrigation into rivers, lakes, groundwater tables, or even seas and oceans. This phenomenon of movement depends, in large part, on climatic conditions, the characteristics of the pesticide in question, topography and cultural practices (Robert 1997). Thus, pesticides can be transferred in two different states: in its dissolved form, or attached to soil particles and moved either horizontally through run-off, generally when rain levels exceed the soil's capacity for absorption, or vertically, through fissures or micropores within the soil (Leonard 1990).

The objective of this study is to evaluate in a specific manner the pesticide residues found in the surface water of the upstream reaches of the Intercontinental Biosphere of the Mediterranean (IBRM), within the context of a consolidation of international efforts and strategies for the protection, conservation, and multiple and sustainable use of natural resources (Goeury 2009).

11.2. Materials and methods

11.2.1. *Materials used*

11.2.1.1. *Area of study*

This study bears on the surface waters of the Gharb region, which is located upstream within the Intercontinental Biosphere of the Mediterranean (IBRM).

Samples were taken from the primary watercourses of the Gharb region. The first sample was taken from the Oued Drader, a second from where the Oued Drader runs into the Merja Zerga lagoon; two samples were taken from the Nador canal, and a fifth and final sample was taken from the Oued Sebou.

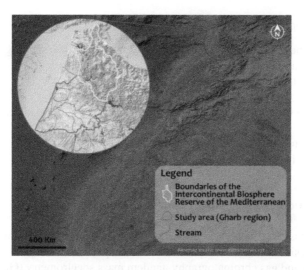

Figure 11.1. *The area of study (Moroccan section) (source: www.elasticterrain.xyz)*

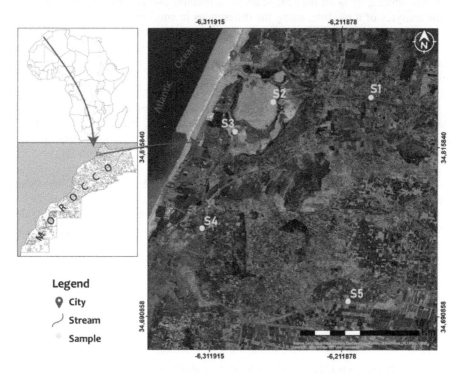

Figure 11.2. *Close-up of the sites from which the analyzed water samples were taken (source: www.elasticterrain.xyz)*

11.2.1.2. Sampling

We used one-time samples taken randomly from the water. During our research, we took five samples of surface water. These samples were taken from three water sources and collected in glass bottles.

The surface water samples were taken from rivers, canals and lakes.

The next step was to conserve these samples in a refrigerator kept below a temperature of 10°C for three days, until their extraction.

11.2.2. Methods used and procedures of analysis

The samples were analyzed in the laboratory through two forms of the most popular analysis technique: liquid chromatography–tandem mass spectrometry (LC–MS/MS) and gas chromatography–tandem mass spectrometry (GC–MS/MS).

Chromatography is an immediate analysis method that allows for the separation of the analytes of a mixture, using the differences in equilibrium constants between the compounds, while they are suspended between a mobile phase in which they are soluble and a "fixed" or stationary phase, which exerts a retarding effect on them (Rosset et al. 1995). This technique comes in a variety of forms depending on the nature of the phases (gaseous, solid or liquid).

Thanks to its speed and sensitivity, gas chromatography is more commonly used than liquid chromatography. However, the potential of the latter is far more important, as approximately 85% of known compounds are not sufficiently volatile or thermally stable enough to undergo gas chromatography (Christian 1993).

LC also has the advantage of increased polyvalency because of its three interacting analytes (solution, mobile phase, stationary phase), as opposed to GC's two (solution and stationary phases), and because of the variety of available stationary phases. Finally, the pairing of this technique with mass spectrometry through sources of atmospheric pressure allows for the overcoming of LC's primary fault, the poor performance of its detectors (Christian 1993).

11.3. Results and discussions

11.3.1. *Pesticide use*

The water analyses carried out in our area of study have allowed us to record 38 different active materials. Organochloride pesticides are the most commonly used (making up 29% of the analyzed pesticide types). Organophosphate pesticides (21%)

were also among the most commonly used of the important chemical classes (organophosphates, organochlorides, pyrethroids, carbamates, organonitrogens). Other pesticides (covering nine chemical families) make up a non-negligible portion of the use of pesticides in this area (Figure 11.3).

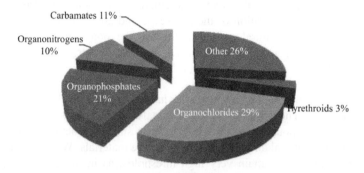

Figure 11.3. *Chemical classifications of pesticides in the upstream reaches of the Intercontinental Biosphere of the Mediterranean (IBRM)*

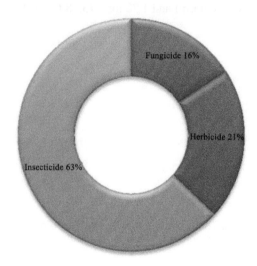

Figure 11.4. *Pesticide classifications according to targets in the upstream reaches of the Intercontinental Biosphere of the Mediterranean (IBRM)*

In terms of targeting, there are different categories of pesticide within the Gharb region, notably insecticides, fungicides and herbicides. Our analyses show the clear predominance of insecticides, followed by herbicides (63% and 21%). Products that act on fungal diseases are less sought after, with fungicides representing 16% of the total products used (Figure 11.4).

11.3.2. Water compartment contamination risks in the upstream reaches of the Intercontinental Biosphere of the Mediterranean (IBRM)

The contamination of surface water by the pesticides used in agriculture is influenced by the natures of different soils, climatic conditions, the structure of the drainage basin, the organization of the landscape and above all by the interactions between different agricultural activities. Indeed, the potential for contamination depends largely on practices (quantities used, application dates, soil work, etc.) and the layout of the landscape (grassy strips, ditches, etc.).

The analyses carried out on our five sites have shown elevated pollution levels at the study stations. The group of pesticides analyzed at these stations far surpass standards for water potability (OMS 1994), those standards being 0.1 µg/L for an individual active material and 0.5 µg/L for all active materials. We can also note that the contamination levels attain very high thresholds, as in the case of the station (S%) on the Sebou river, which registered a number over 43.3. times that required by the OMS. Stations S2 and S1, representing the Drader river and the mouth of the Drader at Merja Zerga, remain the least contaminated, with average concentrations of 2.41 µg/L (4.8 times the norm) and 1.78 µg/L (3.58 times the norm) respectively (Figure 11.5).

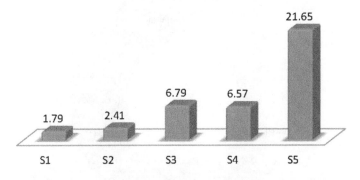

Figure 11.5. *Total concentrations of pesticide residue in the different stations analyzed throughout the area of study (µg/L)*

Therefore, contamination depends on local factors, such as the use of the soil, the agricultural practices chosen by farmers and the paths worn into the land by water.

The average concentrations of active pesticide materials, when analyzed according to chemical family, exceed permitted concentration standards, with the exception of the pyrethroid family (Table 11.1).

	Pesticide levels (µg/L)					
Family	Carbamates	Organophosphates	Organochlorides	Organonitrogens	Pyrethroids	Other
Average level	1.29	1.00	2.52	0.53	0.26	2.25

Table 11.1. *Average pesticide concentrations according to the chemical family in the analyzed stations within the area of study (n = 5 stations)*

Organochlorine pesticides, synthetic aromatic molecules possessing one or two chlorine atoms, have been used in enormous quantities across the globe as contact insecticides, and to a lesser degree as fungicides or acaricides. Their action spectrum is thus very large, and through their repeated use, their efficacy has gradually diminished, leading farmers to increase the frequency and dosage with which they are applied.

Organochlorine pesticides are characterized by their semi-volatility, which makes them extremely mobile. Once evaporated, a compound can cover great distances through masses of air (Abou-Donia et al. 1990). Their persistence in the environment is remarkable, the carbon–chlorine bond within each of their molecules being highly stable. This gives them great resistance to deterioration and an increased half-life, particularly when the chlorine atom is bonded to an aromatic ring (Angerer and Ritter 1997).

Active materials detected	Total pesticide levels (µg/L)				
	Station 1	*Station 2*	*Station 3*	*Station 4*	*Station 5*
Thiamethoxam	–	–	0.49	0.46	–
Aldrine	–	–	–	–	0.8
Dieldrine	–	–	–	–	1.36
HCH-alpha	–	–	–	–	0.26
HCH-beta	–	–	–	–	0.86
Heptachlor	–	–	–	–	1.66
4,4'-DDD	–	–	–	–	1.83
4,4'-DDE	–	–	–	–	1.62
4,4'-DDT	–	–	–	–	1.76
Endosulfan-sulfate	–	–	–	–	0.96
Hexachlorobenzene	–	–	–	–	0.52
Total of organochlorides	0	0	0.49	0.46	11.63

Table 11.2. *Organochloride pesticide concentrations in the analyzed stations within the area of study (n = 5 stations)*

According to our analysis, no organochloride was present in the Drader river. In the Nador canal, a single example of this chemical family was found (thiamethoxam), at an average concentration exceeding 0.47 µg/L. It was in the Sebou river that the aromatic derivatives DDD and DDT were recorded, at elevated concentrations of 1.83 µg/L and 1.76 µg/L respectively.

11.4. Evaluation of the risks of pesticides to human health

The risks of human pesticide poisoning stem from both the active substance's toxicity level (acute or chronic toxicity) and the exposure to the pesticide (the absorbed dose per day; the quantity of residue present). Humans can be exposed to pesticides in several ways: voluntary or involuntary ingestion (unwashed hands), inhalation or direct skin contact (Hartung 2007).

Organochlorine pesticides, the predominant variety in this study, have liposoluble and bioaccumulative properties: after having reached natural spaces, they are in large part stored within living organisms, which must store them because of their resistance to metabolization. They have a tendency to dissolve into lipids, which allows them to easily pass through the phospholipid structure of biological membranes. They accumulate in the fatty tissues of living organisms and are concentrated by passing from one species to another through the food chain until they reach humans (Rochez et al. 2003).

Our water analyses show concentrations exceeding OMS recommendations for pesticide residues, anything over 0.1 µg/L per substance for water used without preliminary dilution (raw water). Furthermore, two active materials (atrazine and simazine) were found, which are banned in Europe, but which remain in use within Morocco.

The direct use of water from these rivers for consumption or for agricultural use (crop irrigation, livestock watering, etc.) thus represents a real danger.

11.5. Evaluation of the risks of pesticides for the environment

The pesticides used as agricultural treatments can undergo several processes (Mathur et al. 2010):

– transformation through metabolization by microorganisms, photolysis or catalysis;

– retention, whether through absorption by vegetation or soil microflora or by absorption into the water content of the soil;

– movement, through leaching, being washed away, or through run-off, all potentially leading to the contamination of drainage, surface or groundwater.

The deterioration process of pesticides can play a major role in the depollution of contaminated environmental areas if it leads to total mineralization; if not, the metabolites created through deterioration can cause pollution in their turn (Multigner 2005).

Thus, diffused pesticide pollution can reach different areas of the environment. Their persistence and mobility through space and time can have negative effects that degrade the ecosystems of the Intercontinental Biosphere of the Mediterranean (IBRM) downstream of our area of study.

11.6. Conclusion

The present study shows us that the level of water contamination by pesticides in the Gharb region, primarily caused by insecticides, is in fact non-negligible. Organochlorides and carbamates are the most widespread chemical families. Boscalid (in the carbamate family) is the fungicide with the highest average levels, reaching 0.82 µg/L. Pirimiphos-methyl is found in small doses compared to other active materials; the total concentration of pesticides by station varied from 1.8 µg/L to 21.6 µg/L, with the highest concentration being found in the Oued Sebou, and the lowest in the Oued Drader.

The methods used to analyze the pesticide residues in the water samples had the capacity to detect their presence even in weak concentrations. A regular tracking program will need to be established in the region to limit the contamination of the environment by toxic substances, and above all to stop the introduction and use of pesticides banned within developed nations such as those of Europe (atrazine and simazine having been detected within the study).

It is thus essential today to enlarge the field of ecosystem protection and conservation, especially in protected areas, and to apply the foundations of sustainable development, moving from ecological engineering to territorial engineering.

In this vein, and with the end of protecting the environment of the Intercontinental Biosphere of the Mediterranean (IBRM), which is the subject of international agreements, it proves crucial to guarantee the involvement of local and peripheral populations in the conservation and preservation of natural reserves, and to stay true to promises to apply the principles of sustainable development to the region.

11.7. References

Abou-Donia, M.B., Nomeir, A.A., Bower, J.H., Makkawy, H.A. (1990). Absorption, distribution, excretion and metabolism of a single oral dose of [14C]tri-o-cresyl phosphate (TOCP) in the male rat. *Toxicology*, 65, 61–74.

Angerer, J. and Ritter, A. (1997). Determination of metabolites of pyrethroids in human urine using solid-phase extraction and gas chromatography-mass spectrometry. *Journal of Chromatography B: Biomedical Sciences and Applications*, 695(2), 217–226.

Christian, G.D. (1993). *Analytical Chemistry*. John Wiley & Sons, New York.

Comité de la prévention et la précaution (2006). Risques sanitaires liés à l'utilisation des produits phytosanitaires. Report, Comité de la prévention et la précaution.

FAO (2002). Réduction des risques pour la santé et l'environnement. Report, United Nations.

FAO (2010). Code international de conduite pour la distribution et l'utilisation des pesticides de la Food and Agriculture. Report, FAO, Rome.

FAO and OMS (1957). Méthodes d'essai toxicologique des additifs alimentaires : deuxième rapport du Comité mixte FAO/OMS d'experts des additifs alimentaires. Report, FAO and OMS, Geneva.

FAO and OMS (1995). Évaluation des risques dus à la présence de produits chimiques dans les aliments. Issu de l'analyse des risques dans le domaine des normes alimentaires, par le Comité mixte FAO/OMS d'experts des Normes alimentaires. Report, FAO and OMS, Geneva.

Gasnier, C., Dumont, C., Benachour, N., Clair, C., Chagnon, M.C., Séralini, G.E. (2009). Glyphosate-based herbicides are toxic and endocrine disruptors in human cell lines. *Toxicology*, 262(3), 184–191.

Goeury, D. (2009). Protéger la réserve de la biosphère intercontinentale de la Méditerranée Andalousie (Espagne) – Maroc. In *Mers, détroits et littoraux : charnières ou frontières des territoires ?*, Semmoud, B. (ed.). L'Harmattan, Paris.

Hartung, T. (2007). ESAC statement on the OECD adopted Test Guidelines for acute oral toxicitytesting. Report, European Commission, European Centre for the Validation of Alternative Methods.

Leonard, R.A. (1990). Movement of pesticides into surface waters. In *Pesticides in the Soil Environment*, Cheng, H.H. (ed.). Society of America Book, Madison.

MARA (1984). Arrêté du ministre d'Agriculture et de Réforme Agraire relatif à l'interdiction des pesticides organochlorés, 466. Legislation, FAO/FAOLEX/ECOLEX, Rabat.

Mathur, N., Pandey, G., Jain, G.C. (2010). Pesticides: A review of the male reproductive toxicity. *Journal of Herbal Medicine and Toxicology*, 4, 1–8.

Multigner, L. (2005). Effets retardés des pesticides sur la Santé Humaine. *Environnement, risques & santé*, 4, 187–194.

OMS (1994). Directive de qualité pour l'eau de boisson. Recommandation, OMS, Geneva.

Pouradier, G. (2010). *Vous reprendrez bien un peu de pesticides ? De graves conséquences sur la santé*. Eyrolles, Paris.

Provost, D., Cantagrel, A., Lebailly, P., Jaffre, A., Loyant, V., Loiseau, H., Vital, A., Brochard, P., Baldi, I. (2007). Brain tumours and exposure to pesticides: A case-control study in south-western France. *Occupational and Environmental Medicine*, 64, 509–514.

Ramade, F. (1990). La Conservation des écosystèmes méditerranéens. Les fascicules du Plan Bleu. *Economica*, 3.

Ramade, F. (ed.) (1996). La conservation de la Nature en Méditerranée. In *Éléments d'écologie, écologie appliquée*. Ediscience International, Cachan.

Ramade, F. and Vicente, N. (1994). La conservation des écosystèmes côtiers et marins méditerranéens. In *Les biocénoses marines et littorales de Méditerranée : synthèse, menaces et perspectives*, Bellan-Santini, D., Lacaze, J.-C., Poizat, C. (eds). Éditions du Musée national d'histoire naturelle, Paris.

Robert, M. (1997). Dégradation de la qualité des sols : risque pour la santé et l'environnement. *Bulle. Acad. Natel. Med.*, 1(21), 42.

Rochez, H., Buet, A., Tidou, A., Ramade, F. (2003). Contamination du peuplement de poissons d'un étang de la réserve naturelle nationale de Camargue, le Vaccarès, par des polluants organiques persistants. *Revue d'Écologie*, 58(1), 77.

Rosset, R., Caude, M., Jardy, A. (1995). *Manuel pratique de chromatographie en phase liquide*. Masson, Paris.

Sapozhnikova, Y., Bawardi, O., Schlenk, D. (2004). Pesticides and PCBs in sediments and fishfrom the Saton Sea, California, USA. *Chemosphere*, 55, 797–809.

12

Forest Fires: Their Impact on the Sustainable Development of the IBRM

12.1. Introduction

The conservation and sustainable development of natural resources is a means towards and a major challenge for ensuring human subsistence. Forest ecosystems represent a cluster of capital, alimentary, material, medicinal and combustible resources, among other kinds. Thanks to its ability and capacity for natural (without human intervention) or artificial (with human intervention) regeneration, forests offer a range of natural biodegradable products which respond to energy, equipment and infrastructure needs.

Forests, today subject to overwhelming pressures, also contribute to the regulation of the water cycle and the conservation of soils, which are able to absorb significant quantities of carbon emitted through human activities and to host a significant biodiversity.

Summits, meetings, treaties, conventions and projects have been held in an attempt to organize societies towards more sustainable forms of human development, setting 17 sustainable development goals for 2030, divided into 169 targets that must be achieved before that date.

One particularity of the Mediterranean region is the vulnerability of its forest ecosystems to fires. Forest fires represent a serious threat to their durability, differing according to the social context of each area and the impact of the use of forest spaces by each community.

Chapter written by Rachid SAMMOUDI, Abdelkader CHAHLAOUI, Nadia MACHOURI, Lahoucine AMZIL, El Habib EL AZZOUZI, Reda NACER, Kawtar JABER and Maya KOUZAIHA.

Figure 12.1. *Site of Biological Importance in the Moroccan section of the IBRM (source: M. Derak)*

In the Moroccan section of the Intercontinental Biosphere of the Mediterranean (IBRM), forest fires are set to open up new spaces for agricultural use. There are large areas expanded for the illegal growth of cannabis, developed by users on specific plots of land, and in forest spaces where the natural vegetative cover has been cleared and burned, complicating the question of the public ownership of forests, and aggravating conflicts between users of the space and supervisory bodies. The region is thus one of tensions linked to the expansion of cannabis crops: users escaping the control of the authorities, to sell the crop and the final products of the process. The burning of millions of hectares of forest each year degrades natural spaces and their ecological roles, particularly in the Chefchaouen, Larache and Tétouan provinces.

These fires can degrade the natural heritage of the Rif region, and particularly the forest ecosystems of the Chefchaouen province and its surroundings, where the planting of cannabis continues, representing a major challenge for the Moroccan government's implementation of the SDGs.

How can a general ecosystemic equilibrium be established and conserved, and how can the private ownership of forest ecosystems in particular be secured? How can local development be promoted? How can village populations be accompanied, so as to ensure that they develop renewable resources and rationalize their industries, so as to diversify their sources of revenue?

12.2. The phenomenon of forest fires in the northern provinces

The perception of forest fires changes from one ecological region to another and from one country to another, depending on the scale of their propagation, the context and origin of outbreaks, the positioning of the relevant actors (scientists, whether ecologists or not, managers, politicians, etc.) and the strategic direction chosen for the management of the forest space in question.

While natural fire is an indispensable part of the life cycle of forests, its arrival and dispersal are also considered to be a disaster for forest ecosystems, spaces close to inhabited areas and infrastructures, and a threat to life and limb through the toxicity of the smoke they create. Fires ravage forest spaces every summer, be they natural (set off by lightning or through a combination of dryness and intense heat), criminal or accidental.

In the absence of pastoral activity on the northern shores of the Mediterranean, some millions of hectares have gone up in smoke through these last few years in France, Portugal, Spain and Greece. In comparison, forest fires remained relatively few in number and superficial in scale on the southern shores. They nevertheless represent a serious threat to the biodiversity of the northern Moroccan Rif region, where they are more frequent and more drawn out in comparison to those across the rest of the country. Vulnerability to fires is significant from June to October, with a Fire Weather Index (FWI) ranging from 30 to 55 (Jesus et al. 2019), influenced by a dry spell increasing the flammability of the biomass, the exposed and rugged terrain, and by the consequent inaccessibility of the massifs, which hampers the speed with which the fires can be fought.

In this area, more than 80% of fires take place in August and September, and the majority are criminal in origin. The average number of fires started and the average area they cover over the last decade have reached 39 fires and 497 ha per year, with a high of 1,371 fires recorded in 2019 (Figures 12.1 and 12.2). Investigations led by local forestry authorities into the causes of the fires concluded that almost all (99%) of the fires resulted from human interventions, whether through imprudence or deliberately, so as to clear forest spaces for the growing of crops (DPEFLCD-CH 2019).

The extent of the fires has increased over the years where there is significant biomass built up through humid winters and springs (the development of grassy and shrubby strata), constituting an abundance of combustible material for the dry season, which promotes the burning of the arboreal stratum. Their severity depends on the density of the understory, the presence of particularly flammable species and the humid, sub-humid and semi-arid bioclimates that dominate around 60% of the region (Mharzi Alaoui et al. 2015), these phenomena being amplified by the rugged and exposed terrain and its inaccessibility.

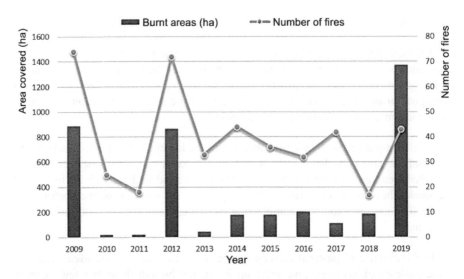

Figure 12.2. *Change in the number of and area covered by forest fires declared within Chefchaouen province 2009–2019 (source: DPELCD Chefchaouen). For a color version of this figure, see www.iste.co.uk/romagny/biosphere2.zip*

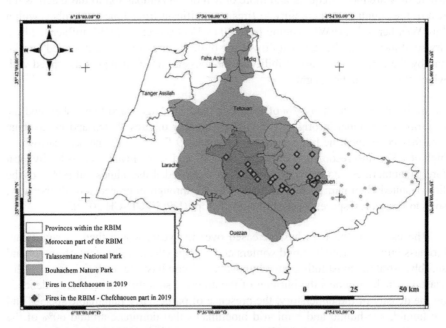

Figure 12.3. *Locations of forest fires in Chefchaouen province in 2019. For a color version of this figure, see www.iste.co.uk/romagny/biosphere2.zip*

12.3. Links between sustainable development and forest fires

Chefchaouen province is made up of approximately 50% of forest space and wasteland, 5% of urbanized areas and 45% of hilly agricultural land, little suited to modern farming and so assuring sustained revenue for the farming population.

The population is 90% rural, with a density of 129 habitants/km^2, which is higher than the rest of Morocco (HCP 2014). Village activities are essentially based around food-producing crops (grains, fruit trees, particularly figs and olives) grown on sloping plots of private land near the forest, and around the farmyard breeding and the maintenance of small family herds made up primarily of goats.

The Utilized Agricultural Area (UAA), insufficient for productive farming, increases anthropogenic pressure on the forest. Through the last two years, cannabis growth has become the main activity of over 90% of village populations, including those without land (ONUDC 2005). It represents more than 60% of the agricultural production of Chefchaouen province and seems to take up over 40% of the UAA, unmentioned within official statistics (Lakhouaja et al. 2017).

The degradation of the vegetative cover exacerbates the impoverishment of the soil by cannabis monocultures, water erosion, the undermining of riverbanks and the development of badlands. Slash-and-burn agriculture is also a classic means of fertilization for cannabis growth.

Despite a provincial herd on the order of 280,000 heads of smallstock, 50% of which are goats (HCP 2017), their extensive grazing remains nonaggressive with regard to forest resources when compared to other areas with the same biogeographical conditions, notably the Mid-Atlas mountains where the herbaceous stratum becomes rare or absent in summer.

The rugged terrain limits the movements of herds through forested massifs, and the size of a family herd does not exceed a dozen heads. This size means that most households cannot have their herd guarded by a family member. The women of the house gather fodder from forest spaces by trimming young shoots from the arboreal stratum (mainly deciduous trees) as well as gathering firewood (brush and shrubs) to meet their households' fuel needs.

Commercial logging offences are almost entirely absent, and the legal gathering of firewood (from loose deadwood) by the local population, enshrined in rights of use, seems insignificant with regard to reducing the risk of forest fires.

Faced with the fragility of ecosystems, an ecological disequilibrium that continues to grow, a socio-economic context equally out of balance and pressure from international bodies, the Moroccan state has undertaken several projects with the aim of reconciling local development, the conservation and protection of forest ecosystems, biodiversity, the public ownership of forests and the demands of international bodies.

The Moroccan government has initiated several integrated development projects since: the 1960s with DERRO, with the World Bank; the 1990s with GEFRIF; the 2000s with MEDA Chefchaouen; DRI-GRN in collaboration with the EU; INDH phases I and II from 2005; Green Morocco; and multi-sector projects (the promotion of ecotourism and artisanship, an antislavery program, the fight against the cold wave, etc.). It also created an Agency for the Development of the Northern Provinces (ADPN) to promote a dynamic of territorial development and promotion, and to improve the unity of economic and social projects and programs. These packets of investments and measures were deployed to boost the local economy, limit the spread of cannabis growth, better promote the area's natural resources and the potential of the landscape and its climate, protect forest ecosystems, reduce the rural exodus and immigration, and ensure social peace.

12.4. Conclusion

The analysis of the causes of forest fires reveals that biophysical factors (the flammability of local species, the climate, the topography, exposure and inaccessibility) have not alone induced the degradation of forest resources and biodiversity. Socio-economic factors represent the primary cause of forest fires in the Moroccan section of the IBRM, which is being used to gain usable space for growing cannabis. The strong rural demography leads to a dependence on the part of the riverside population on forest resources and nearby publicly owned land for its subsistence, this being in the absence of land for cultivation (small, elevated plots) and other revenue-generating activities, with the exception of the illegal growth of cannabis.

The legalization of cannabis growth, held to be illegal since the 1960s, the withdrawal of cannabis resin from the list of drugs and the accounting for the resulting income within Morocco's GDP would be a gift and an economic opportunity. In this hypothesis, it would be essential to put in place measures for securing the forest and forest resources, establishing sustainable management in accord with the population, promoting forest resources and the landscape, promoting social protection and reducing regional inequalities, revisiting the urban-oriented policies of the province through the creation of centers or satellite towns structured so as to modernize the way of life of villagers, and banning sovereign dependence on natural resources and free access.

The protection, security and promotion of natural ecosystems will not be possible without the effective involvement of the local population and local authorities under the aegis of the managing authority of the new "Moroccan Forests 2020–2030" strategy.

12.5. References

Chouvy, P.A. (2018). From kif to hashish. The evolution of the cannabis industry in Morocco. *Bulletin d'association de géographes français*, 95(2), 308–321. doi: 10.4000/bagf.3337.

DPEFLCD-CH (2019). Rapport annuel des incendies de forêt de la direction provinciale des eaux et forêts et de la lutte contre la désertification de Chefchaouen. Report, DPEFLCD-CH.

Goeury, D. and Semmoud, B. (2009). Protéger ou contrôler le détroit ? La réserve de la biosphère intercontinentale de la Méditerranée Andalousie (Espagne) – Maroc. In *Mers, détroits et littoraux : charnières ou frontières des territoires ?*, Semmoud, B. (ed.). L'Harmattan, Paris.

HCP (2014). Rapport National de recensement général du Maroc. Report, HCP.

HCP (2017). Annuaire Statistique de la Région Tanget Tétouan AL Hoceima. Report, HCP.

Jesus, S.M.A., Tracy, D., Roberto, B., Giorgio, L., Alfredo, B., Daniele, D.R., Davide, F., Pieralberto, M., Tomas, A.V., Hugo, C. et al. (2019). Advance EFFIS report on forest fires in Europe, Middle East and North Africa 2018. Report, Joint Research Center. doi:10.2760/262459.

Lakhouaja, E.H., Faleh, A., Chaaouan, J. (2017). Cannabiculture et feux de forêts dans la province de Chefchaouen : analyse cartographique et statistique. *Revue Tidghine des Recherches Amazigh et du développement*, 6, 14–34.

Mharzi Alaoui, H., Assali, F., Rouchdi, M., Lahssini, S., Tahiri, D. (2015). Analyse de l'interaction entre l'éclosion des feux de forêts et les types de bioclimat au Nord du Maroc – Cas de la région du Rif occidental. *Revue Marocaine des sciences agronomiques et vétérinaires*, 3(3), 46–53.

ONUDC (2005). Enquête sur le cannabis 2004. Report, ONUDC.

13

The Social and Solidarity Economy and Biodiversity in the Intercontinental Biosphere of the Mediterranean

13.1. Some framing of the concept of the social and solidarity economy

From the start, the application of nature protection policies has been a source of controversy, the foundations of which extend far beyond the evident opposition between the economic logic of development and the ecological logic of conservation. As a management apparatus, the social and solidarity economy (SSE) approach designates a cluster of economic initiatives with social aims that help construct new ways of living and thinking. It places the human being at the center of economic and social development, privileging principles of equality, equity, solidarity and the communal management of resources. The SSE aids the creation of wealth and employment and therefore of sustainable economic development. The relationship of the SSE to lands and spaces refers to several different dimensions with regard to the environment, in the primary sense of the word (that which surround on all sides): (I) a social space (civil environment); (II) an ecological space (natural environment); (III) a space of organization (practical environment); (IV) a space of power (political and administrative environment); and finally (V) a space of identity (cultural environment).

From our reading of Muhammed Yunus' 2017 work *A World of Three Zeros (Vers une économie à trois zéros)*, we retained three major propositions for developing "otherwise": (I) the emancipation of young creative people; (II) the

---

For a color version of all the figures in this chapter, see www.iste.co.uk/romagny/biosphere2.zip.

Chapter written by Hicham ATTOUCH, Soukaina BOUZIANI and Sonia ADERGHAL.

liberalization of technology to the end of a sustainable economy; and (III) "society for all", so as to reduce inequalities. In this order of ideas, to develop "otherwise", on a sustainable basis and with the leadership of, among others, social and solidarity economy organizations (SSEOs), demands a new paradigm centered around new relations between humans and nature.

Indeed, for more than two decades, SSEOs, and in particular cooperatives, have demonstrated their capacity for resilience in the face of crises, and for leading projects that help improve the human condition and the development of common properties. At a practical level, cooperatives are flourishing businesses that have made significant contributions to the reduction of poverty, the promotion of gender equality, the provision of healthcare services, the fight against HIV/AIDS and the maintenance of a sustainable environment (Birchall 2004).

We summarize in Figure 13.1 the ingredients used by SSEOs to allow them to contribute to developing "otherwise" through liberalization and the encouragement of citizen initiatives.

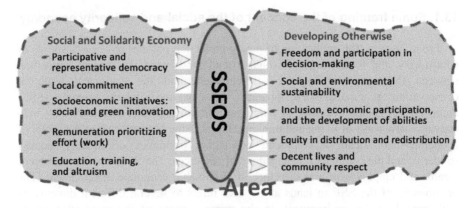

Figure 13.1. *The SSE and sustainable territorial human development (source: Attouch 2011)*

It is worth noting that the term SSE is often defined as an alternative mode of production. In Morocco, the development of the SSE has taken support both from ancient traditional bases and from the international evolution of the concept. It is considered to be less a response to exclusion and the crisis of social ties than to poverty and the marginalization of certain groups of individuals or areas. The challenge today is to go beyond simple responses, urgently formulated in response to precarious working conditions and impoverishment, by coherently and articulately structuring SSEs.

Figure 13.2. *SSE objectives (source: chambre régionale de l'économie sociale et solidaire, Île-de-France 2020)*

13.2. Development of natural resources in the Intercontinental Biosphere Reserve of the Mediterranean (IBRM) and the SSE framework

Our data were collected from presidents and vice-presidents of associations and cooperatives within the region in question, following an interview guide. The analysis displays the diverse profiles of the interviewed leaders.

The results show that the objectives of the associations and cooperatives are numerous. Their primary missions bear on environmental protection (the safeguarding and conservation of natural resources, renewable energy, etc.) and helping the local rural population, particularly women: the improvement of living conditions, literacy and education, the financing of projects, revenue-generating activities (RGAs), etc.

We argue that the viability of projects is supported by the signing of multiple local (village associations, INDH (National Human Development Initiative), ministries, agencies, etc.) and international partnerships, notably with Spain. The revenue sources of some associations and cooperatives come from members or from the sale of their own local products.

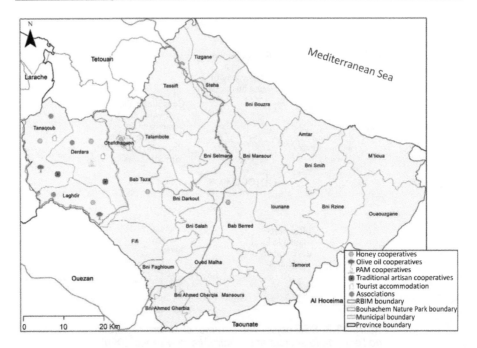

Figure 13.3. *Distribution of cooperatives and associations in the IBRM (source: S. Bouzani 2020)*

The majority of interviewed people suggested that the current state of the IBRM is worrying. Environmental degradation and irrational exploitation of natural resources, above all medicinal plants, are noted.

It appears that the local population carries out numerous agricultural/touristic activities as well as the recycling of waste.

Multiple means have been used to better develop the region's natural resources. Our investigations bring to light the importance of the local population's awareness of the rational use of natural resources through the use of signage, education on respecting the demands of the environment, the maintenance of the infrastructure necessary for water and electricity, etc.

Our analysis of the collected responses allows us to identify the primary problems posed in terms of the environment, specifically fire and dependence on natural resources. It appears that all citizens and all authorities can be seen as stakeholders affected by these different problems.

The solutions envisaged for the area's development can be summarized as follows within a participative logic: local awareness, the launching of RGAs, education, labeling if possible, so as to improve the promotion of local products, and the organization of training sessions to help develop necessary skills, particularly technical skills.

13.3. The role of the SSE in the conservation and development of natural resources

Here we will analyze the responses collected from a certain number of actors in the region. The goal is to explore and highlight the role that the SSE can play in the conservation and development of natural resources in the region.

To assess the value of the SSE's role in the promotion and preservation of natural resources, it seems judicious to adopt a pipeline analysis approach. Economic analysis by pipeline is the analysis of the succession of actions led by the relevant actors to produce, transform, sell and consume a product. This product could be agricultural, industrial, digital, etc. These actions, led successively, simultaneously or complementarily, can be divided into large groups or systems such as production, transformation, commercialization and consumption. Each of these groups encompasses a series of more or less important actions that allow for movement from one to the other, in a logical sequence of interventions: we speak thus of actions situated upstream and downstream within the pipeline. These groups can themselves be split into subgroups.

The responses given by those we interviewed reinforce our decision to adopt this analytic approach.

The development of the SSE in the region is then motivated by several factors. Actors wished to improve the living conditions of the local population (helping rural women, protecting and developing local know-how, etc.), and to improve local abilities (particularly with regard to sustainable and profitable agriculture, etc.). These objectives were realized in large part through training, awareness and education.

Those interviewed assume that the SSE is truly beginning to establish itself in the region. As an example, this is manifested in the production and commercialization surrounding local products (honey, mushrooms, livestock, etc.), all seeking to augment the value created through this production and commercialization.

Associations	Projects
ADL	– Establishment of ecotourism infrastructure in the douars of Jebel Kelti, in the Talassemtane National Park – Improvement of governance in the Chaouen municipality
PRODIVERSA (Spain)	Rural development projects in collaboration with ADL
ATED	Chefchaouen province tourism strategy
Fundación ETEA para el Desarrollo y la Cooperación (Spain)	Chefchaouen province tourism strategy
Catalan Association for Free Time and Culture (ACTLC) (Spain)	– Chefchaouen province tourism strategy (2002–2007) – Reinforcement of the associative network for the stimulation of productive activities in the rural areas of Chefchaouen (2007–2009) – Project for reinforcing governance
Targa Association	Communal plans for the development of rural communes
Mohammed VI Association for the Protection of the Environment	Actions to improve sustainability and environmental protection in the IBRM
PADE – Development Foundation	Rural development projects
Belgian Technical Cooperation (CTB)	Promotion of rural micro-businesses in northern Morocco through the support of the microcredit sector

Table 13.1. *Projects of the associations within the IBRM (source: field study, Bouziani 2020)*

The diversity of natural resources on offer contributes to the emergence of the SSE. Given that natural resources represent an important basis for the associations/cooperative in question, they have established effectual awareness campaigns around how best to use them. Indeed, RGAs are closely linked to their use.

This smooths the way for socio-economic development in the region and contributes to the improvement of living conditions among the local population, as well as promoting employment.

The establishment and reinforcement of the SSE in the region has moved from the creation of sustainable development projects (agriculture, ecotourism, etc.) and of RGAs, the facilitation of certification (notably from the ONSSA (National Office for Food Safety)), financial support and aid to the commercialization of local products (goat's cheeses, etc.).

> Rural poverty and environmental degradation are ancient problems that have always shadowed the future of the Chefchaouen region. Between 1961 and the launch of the DERRO project and today, the terms of the problem of the Rif mountain's underdevelopment remain the same, but the way in which we can approach its overcoming has, on the contrary, changed (Aderghal 2009).
>
> All of the projects established in the region since the 1990s have sought to apply a participative approach, associating local populations with their action programs. In time, this led to the emergence of a civil society indispensable to the success of development projects.
>
> Local civil society has, in fact, several rural development actions to its name. We can cite, for example, the *Association de développement local* (ADL, Association for Local Development) and *Association talassemtane pour l'environnement et le développement* (ATED, Talassemtane Association for the Environment and for Development) or even the cases of structures put in place with Spanish cooperation: *Mujeres en Zona de Conflicto* (MZC, Women of Conflict Zones) and Institute for the Promotion and Support of Development (IPADE).
>
> Furthermore, the project of creating the "Jnan Rif" fruit drying group appears under the auspices of the SA-PAN (Support Sub-Program for the National Action Plan for the Fight against Desertification and the Effects of Drought) program. It aims primarily to create, at the local level, a socio-economic dynamic that could improve the living conditions and revenues of mountain communities; at an environmental level, this project will also help protect the natural environment against erosion through the planting of fruit trees.

Box 13.1. *The "Jnan Rif" cooperative, a participative project built on social development policies*

The establishment of the SSE is seen by those interviewed as a serious alternative in the fight against poverty, the rural exodus, welfare reliance, etc.

The role of the SSE can be seen within the region through associations and cooperatives that consider natural resources as a key element to be profited from as well as protected.

Nevertheless, the SSE sector has faced a certain number of difficulties that have slowed its development in the region. We can particularly note the absence of synergy and cooperation, a lack of skills, illiteracy, a lack of financial support, etc.

The region has seen the intervention of several actors with regard to training and awareness, namely the provincial Department of Agriculture, the INDH, ONG, OCP, the High Commission for Water and Forests and the Fight against Desertification.

Those interviewed proposed the following solutions for the rational use of natural resources: awareness, environmental education, the reissuing of accounts, etc.

13.4. Conclusion

Our analysis of the territorial dynamic created by the "Jnan Rif" group shows its influence on a wide supply area and a commercial area that has reached a national scale. But this dynamic has not touched the local area in terms of improvements in the quality of local products at the level of the douars that would benefit, nor in terms of improvements to the living conditions of the workers, nor, finally, in terms of the development of general living environments. This suggestion is perhaps tied to the lack of a clear strategy on the part of the actors in the projects, who should have been well-defined upstream, so that the project could have its effects on the level of the douars concerned. But we cannot avoid admitting that the social context in which the project was engaged was also marked by two constraints that prevented it from producing its expected effects: on the one hand, a conflict between different categories of worker around the control of the drying group through the body of the cooperative; and on the other hand, the immediate economic effect provoked by the production of cannabis, which led to disinterest on the part of the relevant douars' farmers towards the group.

13.5. References

Aderghal, S. (2009). Valorisation des produits de terroirs et son impact sur la dynamique territoriale : cas de la coopérative Jnan Rif de séchage des figues et pruneaux à Ain Beida. PhD Thesis, Faculté des Lettres et Sciences humaines, Université Mohammed V de Rabat, Rabat.

Attouch, H. (2011). Économie solidaire et développement humain territorial. *REMCOOP, ODCO*, 1, 69–79 [Online]. Available at: https://www.academia.edu.

Birchall, J. (2004). Cooperatives and the millennium development goals. Report, International Labour Office, Geneva.

Martin, A. (2016). Le paradigme coopératif : une matrice philosophique dévoilant l'Homo cooperatus pour une économie renouvelée. PhD Thesis, Université Laval, Quebec [Online]. Available at: https://corpus.ulaval.ca/jspui/bitstream/.

Martin, A. and Lafleur, M. (2009). Le paradigme coopératif et le développement durable : une réponse nécessaire aux enjeux de la société d'aujourd'hui. Report, Conseil Canadien de la coopération et de la mutualité, Institut de recherche et d'Éducation pour les Coopératives et les Mutuelles, Université de Sherbrooke, Sherbrooke.

Yunus, M. (2017). *Vers une économie à trois zéros*. JC Lattès, Paris.

14

The Media Coverage of the Biosphere Reserve: Ambivalence Between the Protection of Nature and the Promotion of Territories. The Case of RBIM

14.1. Introduction

Morocco has adopted a number of strategies related to environmental communication with the aim of increasing people's awareness. Moroccan TV programs illustrate the implementation of media in the protection of the environment. This proves the state's willingness to improve knowledge about the different values of the biosphere reserves.

The aim of this chapter is to investigate the Moroccan environmental policies and the power of communication in the promotion of the biosphere reserve notion. Its main objective is to reveal the roles and functions of media in educational awareness. The chapter will also try to uncover the roles of stakeholders and their contribution to public awareness. It adopts a mixed approach so as to collect people's attitudes and perceptions on the issue. The data is analyzed through the Statistical Package for the Social Sciences (SPSS) and R INTERFACE for multidimensional analysis of texts and questionnaires (IRAMUTEQ).

Given the urgency and the need to maintain the quality of the biosphere and the environment in Morocco, it has become mandatory to develop environmental education and awareness and increase the participation of people in environmental issues in the biosphere reserves. The implementation of territorial communication can mobilize the populations of the different areas so as to maintain their attractiveness.

Chapter written by Lahoucine AMZIL, Yamina EL KIRAT EL ALLAME and Faiza EL MEJJAD.

14.2. Biosphere reserves: general background

The quality of the environment and biodiversity conservation depends on public awareness. The latter is essential for sustainable human development. The protection of the habitat should be a part of all of the countries' strategies. Various countries have been trying to take decisions and make new choices in environmental governance so as to keep the balance between man and the environment. People's awareness of the value of the environment may reduce environmental deterioration. Hence, the need to exploit media as an efficient means to raise people's awareness of the environmental problems and issues and urge them to take action to protect the environment. Biosphere reserves can play an important role in the sustainable development goals and contribute to improving the economic, environmental and social conditions of the local communities. Biosphere reserves offer good ways to protect ecosystems, and encourage the local economies to build up rural areas. Indeed, as UNESCO puts it, biosphere reserves are "learning places for sustainable development". They also "provide local solutions to global challenges[1]" as they play an important role in the conservation of the ecosystems, landscapes and species.

The aim of this section is to give a brief overview of the people's degree of awareness concerning biosphere reserves and their familiarity with the concepts related to them, their meanings and their definitions. In order to find out about the people's understanding of what a biosphere is, the study investigates a group of participants from different educational levels and different cities.

The present study adopts a mixed-method approach making use of both quantitative and qualitative research instruments. This helps us to recognize the influence of social media practices on the development of environmental awareness in biosphere reserves. The use of qualitative data is basically used to analyze the discourse of stakeholders through the language they use and their behavior towards the RBIM. Quantitative data is also used to measure the behavior of the respondents in RBIM and their opinion about the media in relation to the delivery of information about their environment.

Due to the Covid-19 pandemic, the snowball technique was used with the intention and effect to get the help of the respondents in the recruitment of other respondents. The snowball technique was the most effective way to reach respondents due to the constraints that interrupted the fieldwork process, namely the lockdown/confinement, quarantine, lack of transportation, etc.

---

1 UNESCO (2019). Biosphere reserve [Online]. Available at: https://en.unesco.org/biosphere/about [Accessed: July 2012].

Two questionnaires were designed for the research. The data collected from the public included 45 respondents. The aim of this questionnaire was to find out whether people have a prior knowledge of the concept of the biosphere reserve. The second questionnaire was administered to 30 residents in Chefchaouen, Tetouan and Tangier. The purpose of this questionnaire was to determine whether the residents of a biosphere reserve are aware of the fact that they live in a biosphere reserve. The questionnaire also sought to investigate whether the RBIM in northern Morocco is making use of any media and resources.

A field study was conducted in the RBIM to observe different aspects, namely agriculture, the status of women, social economy and the role of media in relation to each aspect.

The stakeholders and participants were contacted by email for both the interview and the questionnaires during the period between August 2020 and October 2020 in a first round and later on by telephone to conclude their collaboration by a visit to the field during the months of October and November for a period of 15 days.

What do you know about biosphere reserves in Morocco?			
	Number	%	Cumulative percentage
A large wooded area	2	4.4	4.4
Areas where there is no human activity	2	4.4	8.9
Forests in which there are species of animals and plants	8	17.8	26.7
Green areas	1	2.2	28.9
Large lands with animals that should not be hunted	1	2.2	31.1
No idea	14	31.1	62.2
Prohibit hunting for the protection of species	1	2.2	64.4
Protected area in which there are species of animals and plants	2	4.4	68.9
Protected areas located in forests	1	2.2	71.1
Protected places to preserve the environment and species	9	20.0	91.1
Subject to special and strict law	4	8.9	100.0
Total	45	100.0	

Table 14.1. *Definition of biosphere reserve by RBIM's respondents. Source: Field survey (El Mejjad 2020)*

The data analysis reveals that the respondents do not differentiate between biosphere reserve, natural park and natural reserves. As Table 14.1 shows, many of the respondents are not aware of what a biosphere reserve is: 31.1% declare that they have no idea what a biosphere is and only provide definitions which are close to the concrete meaning of the biosphere reserves. Indeed, only 20% of the respondents opt for the statement "protected places to preserve the environment and species".

The respondents' awareness and familiarity with the Moroccan biosphere reserves seems to be very low as 33.3% of the respondents claim not to know much about the issue. Only 13.3% of the respondents seem to be familiar with the RBIM, and 10.0% claim to know the RBIM, RBA. These figures reveal the lack of information about the BRs in Morocco and the urgent need to adopt communication strategies to develop people's awareness of them.

	Number	%
Agadir, Souss Massa	1	3.3
Ain Errami	1	3.3
Akchour	2	6.7
Akchour, OumRabii	1	3.3
Chefchaouen	1	3.3
I don't know	10	33.3
MoulayBouselham	1	3.3
Oasis Zagoura	1	3.3
RBIM	4	13.3
RBIM, RBA	3	10.0
Talassemtane	2	6.7
Talassemtane, Bouhachem	1	3.3
Talassemtane, Akchour, sidiAbdelhamid	1	3.3
Tidghin	1	3.3
Total	30	100.0

Table 14.2. *Respondent's awareness of Moroccan biosphere reserves. Source: Field survey (El Mejjad 2020)*

The survey also reveals that most of the respondents cannot differentiate and distinguish between a BR and a natural reserve as 43.3% claim that they do not know what biosphere reserves are. Indeed, most of the definitions they provide are based on their personal perceptions and knowledge about the issue. They are close to the concrete meaning of the concept as they only relate it to nature based on the terms "biosphere" and "reserve".

	Number	%	Cumulative percentage
A place that preserves biodiversity	1	3.3	3.3
A protected area that combines two continents	1	3.3	6.7
A protected area that combines two continents with plant and animal diversity	2	6.7	13.3
A UNESCO territory that respects biodiversity	1	3.3	16.7
Conservation of biodiversity	1	3.3	20.0
Forests and fauna	1	3.3	23.3
I haven't heard about it before	2	6.7	30.0
I don't know	13	43.3	73.3
I know reserves like Bouhachem and Talassemtane	1	3.3	76.7
Living organisms that live in a natural environment	1	3.3	80.0
The atmosphere, soil, animals	1	3.3	83.3
The first time I've heard about the concept	1	3.3	86.7
The living beings	1	3.3	90.0
The protection of nature	1	3.3	93.3
They contribute to sustainable development	1	3.3	96.7
They mean forest reserves	1	3.3	100.0
Total	30	100	

Table 14.3. *Differentiation between biosphere reserves and natural reserves by RBIM's respondents. Source: Field survey (El Mejjad 2020)*

The results of the key questions addressed in the survey related to the means of information that respondents rely on to find the existence of biosphere reserves in Morocco are presented in Table 14.4 and reveal that 35.6% of the respondents do not seem to be aware of the existence of biosphere reserves in Morocco. Those who report to be aware of them (20%) state that they have have heard about them via the Internet and TV, while only 15% claim to have been informed by friends. This reveals how important and urgent it is for the media in Morocco to promote educational awareness, and for biosphere reserves to develop a strategy to improve their visibility.

	Number	%	Cumulative percentage
Accidentally	1	2.2	2.2
Advertisement signage	1	2.2	4.4
Books	1	2.2	6.7
Conference	1	2.2	8.9
Family	4	8.9	17.8
Friends	7	15.6	33.3
I don't know	16	35.6	68.9
I don't remember	2	4.4	73.3
Internet and studies	1	2.2	75.6
Internet and television	9	20.0	95.6
Work	2	4.4	100.0
Total	45	100.0	

Table 14.4. *Means of information about the existence of biosphere reserves in Morocco. Source: Field survey (El Mejjad 2020)*

It should be noted here that the media in general, particularly some of the Moroccan media, have indeed recently shown interest in the environment and environmental issues as societies are becoming more and more aware of the harmful anthropic actions on the environment. This can explain the respondents' awareness of BRs through the media in the survey. Thus, the media have been playing a key role in developing the people's awareness of the environmental issues and spreading consciousness of the need to protect it. This is, however, still very limited and needs more efforts, a clear vision and efficient measures and strategies.

14.3. The media environment around the biosphere reserve

Certain tools such as academic sources, touristic maps, social media and word-of-mouth information seem to have an impact on the environmental awareness of the respondents of RBIM (Figure 14.2). In analyzing the environmental consciousness of the respondents, it is of relevance to look closely at the sources which play a key role in informing the respondents and being the source of their awareness of the environment in general and RBIM in particular. Figure 14.1 shows that 56.7% of the respondents seem to prefer social networks, while only 16.7% of the respondents claim to favor academic sources, 13.3% rely on press as the main means of information and 3.3% of the respondents depend on touristic signs to get information about the RBIM.

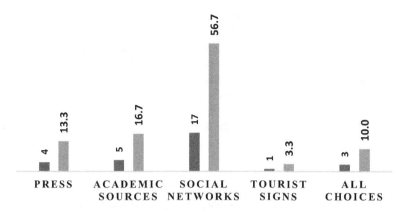

Figure 14.1. *Source of information about the BR/NR.*
Source: Field survey (El Mejjad 2020)

RBIM is a destination that fits in with the international trends in the ecotourism market in Morocco. It is a destination of cultural tourism and authentic experiences close to people and nature. However, some obstacles delay the building of this biosphere reserve image. As Table 14.5 indicates, the respondents complain about the poor infrastructure, the lack of suitable signage in the territory, the distance and the price, and the safety to mention only a few.

	%	Cumulative percentage
Distance and price	11.1	11.1
I don't know	4.4	15.6
Lack of advertising	6.7	22.2
Lack of services	4.4	26.7
Lack of signage	6.7	33.3
Lack of transportation	2.2	35.6
Neglect and lack of concern for natural reserves	6.7	42.2
No problems	8.9	51.1
Pollution and fires	2.2	53.3
Pollution and insecurity	6.7	60.0
Poor infrastructure	15.6	75.6
The outbreak of fires	6.7	82.2
The pollution and deterioration of natural species	8.9	91.1
There are no restrictions	8.9	100.0
Total	100.0	

Table 14.5. *Problems that interrupt the knowledge of RBIM. Source: Field survey (El Mejjad 2020)*

This study also reveals that the respondents do not update themselves on environmental issues and news in their surroundings, which is notable in their answers. Indeed, if a person searches for information concerning the environment in Chefchaouen, Tangier, Akchour and Tetouan on the Internet, they will immediately find the concept of biosphere reserve in their research and educate themselves about it. Actually, this reveals that Moroccans are still not very interested and concerned about the environmental issues and that the idea of cultural or ecotourism is still not widely spread in society. This is partly due to the lack of information about the biosphere reserves in the communication channels as it will appear in the following section.

14.3.1. *Place of the biosphere reserve in the media channel*

Communication channels are powerful and can have an impact on the parties, institutions and society as well. They can be a means for stakeholders, civil society and citizens to cooperate and have a constructive dialogue to address environmental issues in biosphere reserves and try to look for solutions to the challenges facing these areas. Indeed, stakeholders and citizens use different communication channels that can motivate the public participation and develop awareness and collaboration among the members of society. Communication channels can contribute a lot to the development of people's awareness of the environmental degradation and can motivate those who are willing to act to limit environmental degradation and improve and enhance environmental conservation to take action. Thus, people who watch and listen to local stations, that is, MBC 5, 2M and Al Oula, tend to be more attentive to achieve environmental sustainability. Media is the best means for creating this kind of influence. The pictures, videos, plays (Figure 14.2) and documentaries are used to pass the message and spread the information in a simplified form so that everyone can understand it clearly. This game highlights the knowledge and the general culture in the biosphere reserves.

The level of knowledge about the concept of the biosphere reserve in the present study has been identified on the basis of the respondents' attitudes and the way they rate their access to information and news about biosphere reserves. Figure 14.3 shows the numbers and percentages of satisfaction about access to information.

Although the rates for "very satisfied" and "satisfied" seem to be higher than those for "very unsatisfied" and "unsatisfied", the rate for the respondents who are neutral and did not take any position is very high and very close to that of the "very satisfied". The option "neutral" can actually be interpreted in a negative way and as a way to show lack of interest and care about the issue or a way to avoid openly declaring lack of satisfaction.

The type of information the respondents get from the Internet is about fauna, flora, fire, extinct fauna and news related to natural disasters (Figure 14.4). However, other respondents complain about the scarcity of information on television and radio stations. This reveals the necessity to improve the situation and urge the media to become more committed to the cause.

Figure 14.2. *Knowledge game of the biosphere reserve RBIM for students. Source: ATED-Chefchaouen 2016, taken by El Mejjad (2020)*

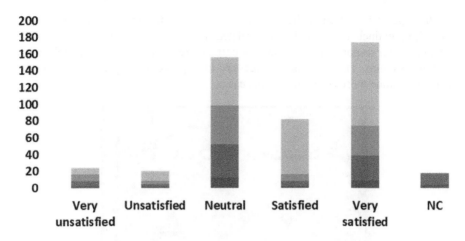

Figure 14.3. *Degree of satisfaction with access to information about NR or BR. Source: Field survey (El Mejjad 2020). For a color version of this figure, see www.iste.co.uk/romagny/biosphere2.zip*

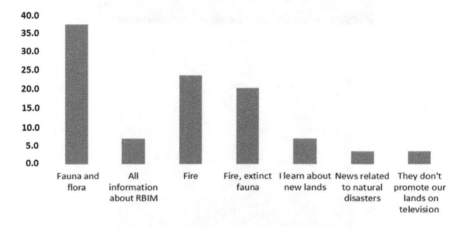

Figure 14.4. *Type of information the respondents get about the environment. Source: Field survey (El Mejjad 2020)*

The respondents were also asked about the information they miss, and their answers were very close and varied between learning about biodiversity and new lands and discovering new reserves in Morocco. These answers show the ineffectiveness of Moroccan programs in providing information about Moroccan lands and landscapes (see Table 14.5).

	Number	%	Cumulative percentage
I want to know more about our country	2	6.7	6.7
Learn about biodiversity	5	16.7	23.3
Learn about fires in forests	1	3.3	26.7
Learn about reserves and natural lands	5	16.7	43.3
Learn about extinct species in our country	5	16.7	60.0
The importance of biosphere reserves	1	3.3	63.3
The location of beautiful reserves	1	3.3	66.7
To discover new touristic lands	1	3.3	70.0
To learn about extinct fauna	1	3.3	73.3
To learn about new reserves	7	23.3	96.7
To learn about new reserves and new cultures	1	3.3	100.0
Total	30	100.0	

Table 14.6. *Needed information according to respondents. Source: Field survey (El Mejjad 2020)*

Table 14.6 shows that the respondents are interested in finding out and learning more about new reserves in Morocco. They also want to learn about biodiversity and natural reserves, and about the extinct species in the country. This is a good indicator of the people's interest in the subject matter and the possibility of improving the degree of environmental awareness in the society.

14.3.2. *Role of media and biosphere reserve actors*

The present study tried to address the issue related to the domains of improvement that should be made in order to improve the attractiveness of biosphere reserves or natural reserves according to the respondents on the basis of statements provided. The highest number of respondents (40%) opted for the statement related to educational awareness and surveillance and the need for more advertising on television. Improving the touristic sector and logistical equipment was highlighted by 26.7% and 22.2% of the respondents respectively (see Table 14.6). This again shows the weak infrastructure and logistics of the touristic sector and the need to improve it.

	Number	%	Cumulative percentage
Cleanliness and safety	4	8.9	8.9
Educational awareness and surveillance, providing more advertising in television	18	40.0	48.9
Improving touristic sector	12	26.7	75.6
Providing logistical equipment	10	22.2	97.8
Management of nature and Earth's biodiversity with the aim of protecting species, their habitats and ecosystems from excessive rates of extinction and erosion of biotic interactions	1	2.2	100.0
Total	45	100.0	

Table 14.7. *Required improvements in BR/NR according to respondents. Source: Field survey (El Mejjad 2020)*

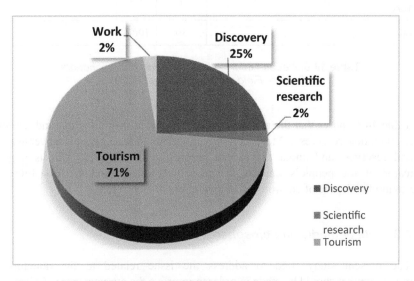

Figure 14.5. *Reasons of visits of RBIM according to respondents. Source: Field survey (El Mejjad 2020). For a color version of this figure, see www.iste.co.uk/romagny/biosphere2.zip*

The promotion of the infrastructure and logistics of the touristic sector will surely promote ecotourism and increase the people's environmental awareness and preservation of biosphere reserves. A number of promotional measures can be used to promote the landscape of RBIM. As Table 14.5 shows, the majority of the respondents, that is, 71%, claim that tourism is the main reason for visiting the RBIM. In fact, the promotion of biosphere reserve by touristic institutions via advertisement, guidebooks and travel agencies can contribute immensely to the mediatization of RBIM.

14.4. Representation of RBIM in the Moroccan media

The representation of biosphere reserves in general and RBIM in particular in the Moroccan media plays a key role in their visibility. The aim of this section is to discuss the role of stakeholders in the visibility and access to RBIM and the measures and strategies to be adopted for improving awareness of the biosphere reserves in Morocco. Each aspect is addressed below.

14.4.1. *Role of stakeholders in the visibility and access to RBIM*

The Department of Water and Forests is working on improving the conditions for the raise of the awareness of the public and their environmental education. One of their main missions, besides the preservation of natural resources and biological wealth in protected areas, is environmental awareness and the education for a large public.

> D.M: […] I would say that access remains at average level. There is work and research on RBIM, unfortunately we do not have specific RBIM websites like European biosphere reserves. […] If you go to the Spanish website, you will find all the information on RBIM. So, the level of communication in Morocco remains average […] (HCEFLCD, Tetouan, 07/11/2020).

This interviewee's testimony is very revealing and shows that even those who work within the system note the stark difference between the Spanish and Moroccan RBIM projects.

Through the national network of protected areas, and by implementing "environmental education programs", the Department of Water and Forests works on establishing an environmental culture among the Moroccan public, spreading awareness regarding the concepts of sustainable development and biodiversity as well as increasing public communication and media debates on these themes. However, despite all of these efforts, the promotion of biosphere reserves in Morocco is still of little impact.

The associations adopt different tools and techniques to promote the RBIM. They also implement many strategies and participate in programs to develop the promotion of the RBIM. However, the ways in which they use the media to approach the public are somewhat outdated.

Another type of media coverage can exist to highlight how associations in Chefchaouen contribute to sustainable human development by improving the living conditions of the population and protecting the environment of the territory of the RBIM.

> [...] The promotion in Chefchaouen is very weak. The ancient cooperatives do not find someone who can help them in this sector. Unlike us we try to make this cooperative well-known and we organize for them training courses to ameliorate their competences in the promotion side [...] (Coordinator, Bouhachem park, 26/10/2020).

The investigation of the issue in North Morocco revealed the presence of a number of institutions in Chefchaouen in charge of the promotion and protection of the biosphere reserves, namely:

– ADL: Association de Développement Local de Chefchaouen (Chefchaouen). The Association for Local Development Morocco "ADL-Al Maghrib", formerly called Association for Local Development in Chefchaouen, "ADL-Chefchaouen" is a nongovernmental organization with a socio-economic and cultural character, but without any profit goals. Its main objectives are to support touristic projects in Chefchaouen, protect the environment, educate, govern and involve the population in the development process (Figure 14.6).

– ATED: Talassemtane Association in Chefchaouen, whose main goals are to spread awareness of environmental issues among citizens, defend the citizen's right to live in a healthy environment, sensitize the citizens, safeguard the environment, carry out programs related to human development in the province, as well as promote ecological tourism.

– Association Assaida AL Horracitoyenneté et égalité: Contribute to the promotion of equity and equality between women and men in the Tangier-Tetouan region.

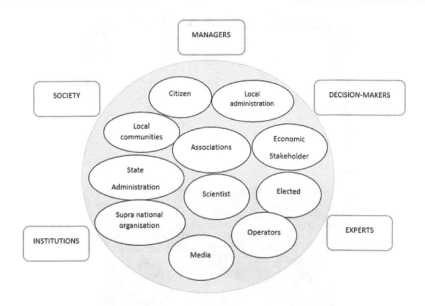

Figure 14.6. *Stakeholders' role in biosphere reserves. Source: UNESCO (2001). Steps and tool towards integrated coastal area management*

Generally, cooperation between the different actors, managers, decision-makers, experts, institutions and society can play an important role in the prevention and control of environmental degradation. It may also help in spreading awareness of new landscapes. It is a difficult task to convince people to come together and achieve a common goal. This requires a lot of efforts and argumentation to bring people together and help all of the members of the community to understand the concept of biosphere reserves, its importance and value. It is crucial to mention here that the local populations are not aware of the potential of their territories.

The study reveals the bad state of the signage of the territory. Besides being weakly developed, it is also hardly visible and very heterogeneous. Therefore, tourists have no indications or points of reference or guidance for reaching their destinations. The existing signage is the result of private initiative. Consequently, there are no concrete projects to develop homogeneous tourist signage on the territory. Finally, the maintenance of this signage is often not assured, and it is usually in a very bad state, as illustrated in Figure 14.7.

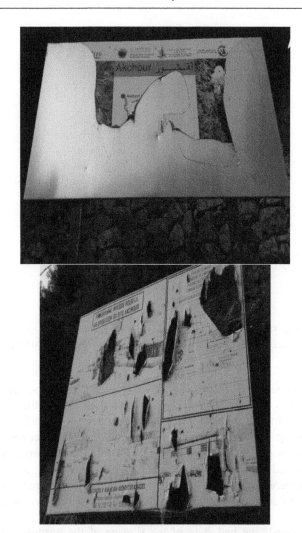

Figure 14.7. *State of the signage in RBIM. Source: Taken by El Mejjad (2020)*

14.4.2. *Measures and strategies for improving the biosphere reserves*

As the previous section reveals, the state of the biosphere reserves needs a lot of improvement at all levels. In an interview with the president of the Talassemtane association, the president declared that they have cooperated with the municipality of Chefchaouen to develop a territorial brand that has been set up under the name "*Gouter Chefchouen*". He stated:

We worked on this brand in a participatory way. We worked with agricultural cooperatives and agricultural products. According to our conception, this territorial brand is a way of producing sustainability in the RBIM.

IRAMUTEQ has been used in textual content analysis and in the description of the textual discourse of the stakeholders in Chefchaouen. Eleven interviews, which were divided into four categories, were conducted in Chefchaouen with Tour guide, members of associations, public institutions and officials in parks and natural reserves. A word cloud was generated, which includes the terms with greater frequency and big size organized graphically. The larger the font and more centered on the cloud, the greater the frequency of the word. This diagram, widely used on the Internet, is a lexicometry tool that simply represents the frequencies of forms by correlating them to their size for the words reserve, biosphere and project (see Figure 14.8). The word cloud shows that other related words such as *natural, product, medium, city, awareness, development, tourism, project* and *information* occur around the term biosphere reserve (see Figure 14.8).

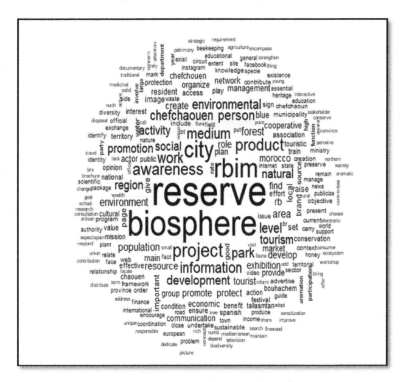

Figure 14.8. *Word cloud generated from the terms biosphere reserve. Source: Field survey (El Mejjad 2020)*

Correspondence analysis, a multivariate graphical technique, is mainly used to explore relationships among categorical variables and is mainly interpreted by comparing the interviewees' statements. The analysis reveals that the park officials and communities hold the same discourse, while the other participants, namely the associations and the engineers of the Water and Forests Department, have a different perspective. The tour guides, for example, tend to talk about tourism and its benefits to the region and its promotion of biosphere reserve via tourism (see Figure 14.9).

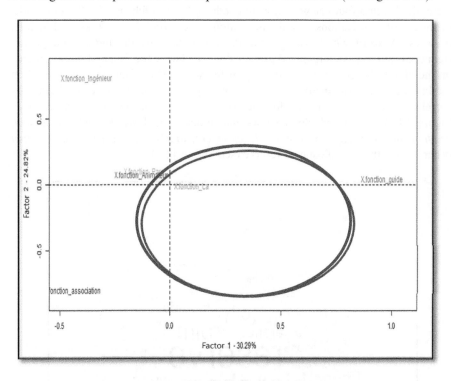

Figure 14.9. *Correspondence analysis and stakeholders' positions. Source: Field survey (El Mejjad 2020). For a color version of this figure, see www.iste.co.uk/romagny/biosphere2.zip*

Figure 14.9 offers a representation of the different stakeholders' positions. It shows the link between the different actors who hold the same discourse and those who have different stands.

The people who contributed to the development of the RBIM are a group of actors who work for the municipality of Chefchaouen, state delegations of agriculture, crafts and tourism, etc. Furthermore, this group of actors are economic actors: producers, cooperatives, restaurateurs, owners, craftsmen, etc.

14.5. Concluding remarks

This chapter has investigated the role of Moroccan policies in the development of communication about the biosphere reserve notion. Its main goal is to determine how the media can contribute to the increase of educational awareness. This study has also looked at the roles of stakeholders and their contribution to public awareness. Given the constraints imposed by the pandemic, the study has undertaken a pilot study to gain a first impression on the issue. As limited as it was, this first investigation has allowed us to generate a word cloud from the terms biosphere reserve, which reveals to be quite interesting.

The main study, which concerned the role of media coverage and communication around the biosphere in general and media coverage of biosphere reserve as a means to achieve the SDGs, did not only reveal that most of the participants are not familiar with the term biosphere reserves but also showed that majority of them do not distinguish between biosphere reserve, natural park or natural reserves. They provide only basic definitions that they understand from the term itself.

This study also unveiled that people are not aware of the social and economic changes in their surroundings and that the respondents' answers vary according to their economic and social background.

The study disclosed the power of communication channels and their impact on both institutions and society at large. In fact, it confirmed that communication can help stakeholders and citizens to cooperate and start a constructive dialogue on environmental issues in biosphere reserves and find solutions to all of the challenges and issues faced. Although stakeholders and citizens use different communication channels, they can both contribute to the implementation of public participation, awareness and collaboration among the public.

The study reported the respondents' complaints about the scarcity of information on the television and radio stations. Indeed, their answers showed the ineffectiveness of Moroccan programs in providing information about Moroccan lands and landscapes.

Among the main measures that could improve the attractiveness of biosphere reserves or natural reserves, the respondents highlighted the need to increase the safety measures, the provision of logistics and equipment, and most importantly the raising of awareness via television.

The study also revealed the local populations' lack of awareness of the potential of their territory. This highlights the importance of developing the required tools for interpreting heritage resources, which would make it possible to educate the population about the resources of their region.

The visibility and homogeneity of the signage of the territory is crucial. Hence, the urgent need for concrete projects is to develop and improve tourist signage on the territory and to guarantee their protection and maintenance in order to promote the ecotourism potentials and resources presented by the RBIM, which has become a destination that is at the level of international trends in the tourism market in Morocco.

Despite all of its limitations, this modest study uncovers the strengths and weaknesses of the biosphere sector and calls for a more exhaustive study that can expand to other areas in Morocco and even to comparative studies with the situation in other countries such as Lebanon, which has a lot in common with the Moroccan context, especially at the cultural level.

14.6. References

Baruah, T.D. (2012). Effectiveness of social media as a tool of communication. *International Journal of Scientific and Research Publications*, 2(5), 1–10 [Online]. Available at: http://www.ijsrp.org/research_paper_may2012/ijsrp-may-2012-24.pdf.

Das, J., Bacon, W., Zaman, A. (2009). Covering the environmental issues and global warming in delta land: A study of three newspapers. *Pacific Journalism Review*, 15(2), 10–33 [Online]. Available at: https://doi.org/10.24135/pjr.v15i2.982.

David Cooper, H. and Noonan-Mooney, K. (2013). Convention on biological diversity. *Encyclopedia of Biodiversity* [Online]. Available at: https://doi.org/10.1016/B978-0-12-384719-5.00418-4.

Gould, J. (2001). Variables in research designs. *Concise Handbook of Experimental Methods for the Behavioral and Biological Sciences*, 3(4), 75–110 [Online]. Available at: https://doi.org/10.1201/9781420040869.ch4.

Herring, S. (2003). Media and language change: Introduction. *Journal of Historical Pragmatics*, 4(1), 1–17 [Online]. Available at: https://doi.org/10.1075/jhp.4.1.02her.

Idelhadj, A. (2000). La Réserve de la Biosphère Intercontinentale de la Méditerranée. Un model de coopération transfrontalière. PhD Thesis, Université Abdelmalk Essaadi, Morocco.

Ishwaran, N., Persic, A., Tri, N.H. (2008). Concept and practice: The case of UNESCO biosphere reserves. *International Journal of Environment and Sustainable Development*, 7(2), 118–131 [Online]. Available at: https://doi.org/10.1504/IJESD.2008.018358.

IUCN (2012). Atelier Régional sur la Gouvernance des Réserves de Biosphère au Maghreb : état et perspectives [Online]. Available at: https://www.iucn.org/downloads/rapport_atelier_gouvernancereserve_de_biosphere_tanger_mars12.pdf.

Johnson, R.B. and Onwuegbuzie, A.J. (2004). Mixed methods research: A research paradigm whose time has come. *Educational Researcher*, 33(7), 14–26 [Online]. Available at: https://doi.org/10.3102/0013189X033007014.

Kaplan, A.M. and Haenlein, M. (2010). Users of the world, unite! The challenges and opportunities of social media [Online]. Available at: https://doi.org/10.1016/j.bushor.2009.09.003.

Keeble, B.R. (1988). The Brundtland report: "Our common future". *Medicine and War*, 4(1), 17–25 [Online]. Available at: https://doi.org/10.1080/07488008808408783.

Krätzig, S. and Warren-Kretzschmar, B. (2014). Using interactive web tools in environmental planning to improve communication about sustainable development. *Sustainability (Switzerland)*, 6(1), 236–250 [Online]. Available at: https://doi.org/10.3390/su6010236.

Livingstone, S. (2009). On the mediation of everything. *Journal of Communication*, 59(1), 1–18.

Livingstone, S. (2011). On the mediation of everything: ICA presidential address 2008 [Online]. Available at: https://doi.org/10.1111/j.1460-2466.2008.01401.x.

Ozyavuz, M., Korkut, A.B., Etli, B. (2006). The concept of biosphere reserve and the biosphere reserve area studies in Turkey. *Journal of Environmental Protection and Ecology*, 3, 638–646.

Project, G.T.Z., Reserve, A.C.N., Reserve, S.B., Abdallah, P., Mobilities, R., Manager, P., Hani, N., Assistant, P., Abou, W., Technical, D. et al. (2010). Shouf biosphere reserve marketing & business plan for Shouf biosphere reserve rural products. Report, Al Shouf Cedar Nature Reserve, Ain Zhalta.

Sadler, G.R., Lee, H.C., Lim, R.S.H., Fullerton, J. (2010). Recruitment of hard-to-reach population subgroups via adaptations of the snowball sampling strategy. *Nursing and Health Sciences*, 12(3), 369–374 [Online]. Available at: https://doi.org/10.1111/j.1442-2018.2010.00541.x.

Schuett, J.L. (2011). Effects of social networks and media on pro-environment behavior. PhD Thesis, University of North Texas, Denton.

Ushanova, I.A. (2015). Mediatization of communication: From concept to theory [Online]. Available at: https://doi.org/10.17516/1997-1370-2015-8-11-2703-2712.Research.

15

Mid-Atlas Cedar Forests and Climate Change

15.1. Introduction

In Morocco, forests occupy a remarkable place among the ecosystems of the Mediterranean, as much for their strong biodiversity as for the ecological and environmental services they provide. Furthermore, national socio-economic choices are closely tied to the climate and its fluctuations. The national economy is highly dependent on water, tourism and the coast, and agriculture and forests, sectors directly exposed to the risks associated with climate change.

The term "climate change" describes any change in the climate over time, whether due to natural variability or provoked by human activities (GIEC 1997). The United Nations Framework Convention on Climate Change refers in its definition of this term only to what can be directly or indirectly attributed to human activities that change the composition of the global atmosphere.

In its evaluation reports, the IPCC presents the possible impact of climate change (scenarios) on water resources, agriculture, ecosystems, coastal and marine areas, human health, and financial infrastructure and services (GIEC 1997, 2000, 2001, 2007a, 2007b). These works describe the progress achieved in our scientific understanding of the human and natural causes of climate change, of observable changes to the climate, climatic processes and their role in this change and the predictions of future climate change produced through simulations. They show that North African countries, especially their rural areas, are particularly vulnerable.

Chapter written by Driss CHAHHOU.

Generally speaking, the regular increase in greenhouse gas emissions, particularly CO_2, is the source of a significant increase in air temperature across the surface of the Earth and of possible climatic changes with likely consequences on forest vegetation and silvicultural practices.

The increase in CO_2 concentrations can produce diverse and concerning effects, particularly regarding changes to forest biomasses, tree/insect and tree/fungus/pest relationships, blooming processes, the fruition and regeneration of forest stands: changes to individual species' distribution areas, to landscapes and strong soil erosion are anticipated, etc. There will also be economic and social consequences: an increased risk of forest fires, the exploitation of wood in the short term and deforestation in the long term, the loss of amenities (leisure activities, tourism, hunting, etc.); populations will be affected with increasing severity by a lack of water and pasture for their herds. In the context of Morocco, this damage to natural and forest spaces is already a reality. Several consecutive years of drought have weakened trees, making them more sensitive to diseases and pathogens. This has led to more or less widespread mortality among certain species, and more or less intense dieback among many others. This is notably the case for the cedar tree (*Cedrus atlantica* Man.) in the Mid-Atlas region.

15.2. General overview of climatic changes

15.2.1. *General aspects of climate change in the Mediterranean region*

The scientific community agrees that the Mediterranean region (including North Africa) will be particularly affected by the disruption tied to climate change. Under this influence, the distribution areas and productivity of forest trees, and Mediterranean ecosystems, are changing (Moreno and Oechel 1995; Déqué 2000). The region is already subject to strong climatic constraints in the summer, when a lack of water and high temperatures appear in tandem. However, climatic models predict that it is precisely in this region that climatic warming will reach its peak in the summer, with a very significant elongation of the dry season, a significant increase in temperature and a decrease in precipitation (Houghton et al. 1996; Gibelin and Déqué 2003). These models foresee a 2–2.5°C increase in temperature over a 30-year period, which could end, in the last of these scenarios, with a notable extension of the Mediterranean region itself (Médail and Ouézel 1996). The limits of vegetation could be displaced towards the north in the Mediterranean zone and upwards in the mountains. The displacement process of such vegetative areas, already underway, will likely extend between 150 km and 550 km northward and from 150 m to 550 m in altitude.

The expected impact on the functioning of ecosystems will be more important in climates of a marked seasonality, where vegetation is subject to strong seasonal climatic constraints (Hoff and Rambal 1999). Indeed, an increased evaporative demand, caused by increases in temperature alongside a decrease in the quantity of available water, will aggravate summer droughts (Rambal and Debussche 1995; Tessier 1999). It is thus essential to understand these mechanisms and their effects on trees and forest stands.

15.2.2. Effects of drought on trees and forest stands

The concept of drought is relative and relies on the existence of a rain deficit. It is an abnormal decrease in rainfall in certain periods of the year defined in relation to a reference or to a climatic average (Brocher 1977). This definition remains incomplete, as it does not take into account the intensity of the constraints imposed on vegetation and its relationship with water availability within the soil. We can define drought as: "a progressive decrease in soil water reserves which, beyond a certain threshold, affects the exchange of water and of carbon, and the development and growth of trees"; we speak in this case of edaphic drought (Brocher 1977; Dreyer et al. 1992).

The first factor responsible for the reduction of water is insufficient rainfall. The trees themselves are the second factor. This second reality often eludes reflections on the matter. In a general fashion (ONF 1999), reductions in growth and in the exchange of gases are observable when the soil water reserves usable by the trees have been reduced to less than 40% of the maximum usable reserve (this being MUR < 0.4). This "lack of water in the soil" can be aggravated by potentially elevated evapotranspiration. Climatic evapotranspiration (Tessier 1999), called potential evapotranspiration (PET), refers to the evapotranspiration of a closed vegetative cover perfectly supplied with water. Real evapotranspiration (RET) depends on the availability of water in the soil and is inferior or equal to PET. Hydric and photosynthetic functions and the growth of trees are favored by an RET/PET near to 1. In the current climatic context, this relation is often below 1, even in areas amenable to forests.

When a period of soil water deficit appears (Granier et al. 1995, 1999), tree water consumption gradually diminishes through a mechanism of localized regulation at the level of leaves, thanks to stoma (Landmann et al. 2003). In the case of prolonged drought, xylem cells can reach their functional limits, which can lead to a massive embolism at the level of roots and sap-carrying xylem cells and to the impossibility of building up reserves (Landmann et al. 2003). These symptoms are accompanied by the rapid dehydration of tissue, even the complete dieback of the tree.

The second factor determinant of the maximum water consumption of a forest stand is its leaf area index, meaning the leaf surface of all of the trees in relation to the ground surface (Breda 1999). The higher the basal area of the stand, the stronger its leaf area index and thus the stronger its water consumption, even though variations in the structure and species composition of a stand can slightly change this trend (Landmann et al. 2003).

In conclusion, limiting the leaf area of a stand is a radical means of limiting the water consumption of the trees. Control over leaf area through silviculture is also the only way for the forester to limit the risk of critical droughts appearing in stands: it aims to tailor the leaf area index of plots to the maximum soil water reserve given local climatic conditions. The variations to be anticipated at the level of the hydric state of the trees, of photosynthesis and of growth will also depend on the characteristics of the soils in question, particularly their ability to hold water (Aussenac et al. 1995).

15.2.3. *The role of forest stands with regard to the water retention capacity of soils*

To influence the water retention capacity of the soil of a forested massif (Dajoz 1996; Breda et al. 2002), it is necessary to determine the principal factors and how they function. If we leave aside the vegetative cover, which retains at best 4% of the precipitation, everything passes into the soil. There are two main qualities necessary for the improvement of this soil's water retention capacity. It must combine:

– a good level of infiltration, on the surface and deeper below it, through small vertical canals;

– a capacity for lateral storage along these canals thanks to other horizontal canals of smaller diameter which will retain this water as it penetrates the soil;

– a good spatial distribution of roots.

Some soils thus possess a good aptitude for water retention and can rapidly retain large quantities of water. Root systems well spread out over a large area and depth contributes to the tangible improvement of this aptitude, thanks to the horizontal and vertical network of roots and rootlets. Optimal root occupation is tied to the structure of the stand and to the species of which it is comprised. The work from which these statements originate suggests that to improve the water retention aptitude of soils, we need an optimal spatial distribution of roots through the soil, consequently needing a good mix of species and a multistage stand (one made up of trees of all ages) to maintain this quality.

15.2.4. *Potential strategies for facing climate change*

There are two complementary global strategies for responding to climate change: adaptation and mitigation (GIEC 2001). Mitigation (GIEC 2001) aims to act on the causes of climate change. Mitigation measures aim to limit the growth of greenhouse gas concentrations in the atmosphere. Two major options are generally considered with regard to mitigation. The first option involves reducing greenhouse gas emissions across different sectors and reducing deforestation. The second option, often called (forest) carbon sequestration, aims to store a part of the atmosphere's carbon in the biosphere.

Adaptation (GIEC 2001) concerns our responses to the effects of climate change and aims to reduce the vulnerability of ecosystems and societies. It relates to any adjustment, whether passive, reactive or preliminary, which can be adopted so as to compensate for the anticipated, expected or existing harmful effects of climate change.

Evaluating vulnerability and impacts is a preliminary phase for adaptation. This vulnerability refers to the degree by which a system can be deteriorated or harmed by changes to the climate.

15.3. The vulnerability of forests to climate change

The vulnerability of Moroccan forests can be understood through the scale of their deterioration and the rhythm of their deforestation, the appearance and extension of diebacks and the deaths of numerous species, as well as through the characteristics of the country's climatic context.

15.3.1. *The vulnerability of Morocco's climatic context and foreseeable changes*

In Morocco, climate elements vary greatly according to geography; average annual precipitation varies from less than 25 mm in the Sahara to almost 2,000 mm in the Rif and Mid-Atlas regions. The distribution of climatic areas on a territorial scale (Mateuh 2000) assigns 560,000 km^2 to the arid and Saharan zone (78%), 100,000 km^2 to the semi-arid zone (14%) and 50,000 km^2 to the sub-humid and humid zone (7%). The majority of the country is thus situated in the arid zone, characterized by prolonged sunshine and severe droughts. Temperatures tend towards extremes, with cold spells over the short term, as well as heatwaves coming in from the Sahara (reaching up to 45°C). The climate is thus characterized by strong irregularity, as well as the beginning of the humid season and the early arrival of the dry season, which determines whether or not there will be rain in the spring.

Global climatic models come together in estimating a likely warming of the Maghreb region in the order of 2°C–4°C in the 21st century (Agoumi 2005). The initial national communication from Morocco evaluated the warming from 0.6°C to 1.1°C and foresaw a reduction in precipitation of around 4% between 2000 and 2020. These changes could have an impact on the frequency and distribution of extreme climate phenomena, particularly those linked to the water cycle, specifically:

– increasingly frequent storms in the north; an increase in the frequency and intensity of droughts in the south and the east of the country;

– a disruption of the seasonal precipitation signal (fewer days of rain and a less persistent rain than in the winter), accompanied by a reduction in snowfall.

This likely change in climate between 2000 and 2020 has had a significant impact on the water cycle and on the demand for and consumption of water, notably for agriculture and forests. Agoumi (2005) estimated that the average flow of surface and groundwater would fall by 10–15% between 2000 and 2020. A policy for adapting to the new climatic context has been imposed to limit the impact of this threat on possibilities of sustainable development in the country.

15.3.2. *Deterioration, deforestation and transformation of forest habitats*

The National Forest Inventory (1996) takes stock of a strong trend towards the reduction of forests under the simultaneous pressures of multiple different factors: clearing to make way for agriculture, the gathering of wood products at a rate that surpasses biological possibilities, overconsumption of the leaves and fruits of the herbaceous stratum by animals and the spreading of urbanized areas and equipment. The forest surface has thus decreased by 245,350 ha in 10 years, a loss judged to sit at approximately 25,000 ha per year. Table 15.1 provides an overview of the rate of decline in the Rif and in other regions of the country (Mhirit and Benchekroun 2006).

In addition to modifications to the arboreal stratum, we must evoke the transformation in species compositions and forest habitats, which has become a clearly acute phenomenon within our forests. There are many highly remarkable groupings that have almost entirely disappeared today, or are present only residually, particularly in cork oak, green oak and cedar groves (Benabid and Fennane 1994, 1999).

Provinces	Forest decline (ha/year)	Rate of decline (in % of initial area)
Al Hoceima	1,036	25.07
Chefchaouen	1,719	32.17
Tetouan	1,762	49.72
Larache	366	35.95
Agadir	478	2.17
Taounate	718	11.76
Benslimane	309	4.56
Kénitra	1,235	17.13
Essaouira	1,341	12.32

Table 15.1. *Rate of forest disappearances in Morocco (source: Mhirit and Benchekroun (2006))*

15.3.3. *Cedar diebacks: an indicator of climate change*

The appearance of cedar diebacks in the forests of the Mid-Atlas region can be traced back to the summer of 2001. Work on Mid-Atlas cedar diebacks (Et-Tobi and Mhirit 2003; Et-Tobi et al. 2006; Et-Tobi 2006, 2008; HCEFLCD 2006a, 2006b, 2006c; Derrak et al. 2008) have allowed the scale of the phenomenon to be recognized (Table 15.2). The impact of the climate can be seen clearly in the 40% of the total area affected by dieback and death. Intensity nevertheless varies from one massif to the next: very strong in the cedar groves of Senoual (49%) and Aghbalou Laarbi (62%), between 31% and 33% in the Bekrit and Azrou cedar groves, and relatively weak in the Jbel Aoua Sud cedar groves (17.47%).

In summary, we can say that almost half of the area covered by the cedar groves studied has felt the impact of the climate.

REMARK.– The section of the Senoual forest studied covers 4.874 ha belonging to Ifrane province; the section belonging to Khénifra province (1,049 ha) was not examined. For Aghbalou Laarbi, the study was limited to the section of the forest belonging to Ifrane province, which covers an area of 15.201 ha representing around 50% of the total area of the forest massif. For the Azrou forest, the forested surface makes up 13,763 ha, with the difference being comprised of clearings.

Forest	Total area (ha)	Examined area	Area of cedars (ha)	% of cedar grove	Area affected by death and/or dieback	% of cedar grove affected
Aghbalou Laarbi	30,275	15,201	5,338	35.12	3,322	62.23
Senoual	5,923	4,874	3,891	79.83	1,892	48.63
Bekrit	10,346	10,346	6,767	65.41	2,272	33.57
Azrou	17,806	13,763	8,679	63.06	2,700	31.11
Jbel Aoua Sud	7,865	7,865	1,740	22.12	304	17.47
Total	72,215	52,049	26,415	50.75	10,490	39.71

Table 15.2. *Scale of cedar diebacks in the Ifrane region (HCEFLCD 2006a)*

15.4. Potential impacts of climate change on cedar forests

In the case of the cedar, changes in the climate (with regard to precipitation and temperature) influence their distribution area as well as their altitudinal limits by migration, and also the composition and structure of their stands, by acting on the physiology of the trees and their associated vegetation.

15.4.1. *Elements of the Atlas cedar vulnerable to climate change*

Rains and their variability represent, in our climate, the most important factor in the distribution of forest species, as well as their growth (Mhirit 1982, 1994, 1999). In this regard, the poles between which the average amount of rain oscillates are 550 mm and 1,800 mm. The role of maximum and minimum temperatures also determines geographical distribution. The average minimum and maximum temperatures are respectively 3.6°C and 18.2°C. In this context, the Moroccan Atlas cedar (Benabid and Fennane 1999; Mhirit 1999) occupies an average altitudinal position of between 1,600 and 2,400 meters; it appears in cold humid and sub-humid bioclimates and individualizes three types of vegetation zones:

– supra-Mediterranean zone, which is home to sclerophyllous and deciduous oaks;

– inferior Mediterranean mountain zone, which is generally pure and dense or mixed with green oak;

– superior Mediterranean mountain zone, which is sparse and home to species of calcicolous grass and juniper trees.

The first consequence of the bioclimate for silviculture is the importance of the average growing season (AGS) of trees or the growing season. For the Moroccan Atlas cedar (Mhirit 1999), the AGS extends in the Rif and the Atlantic Mid-Atlas regions from mid-April to the beginning of July and corresponds to the months in which temperatures remain above 10°C. The second period extends from the end of September to the end of October. In the continental Mid-Atlas and High-Atlas regions, cedars grow in a single phase from mid-April to the end of May; climatic conditions in these regions do not allow for a second phase of growth. This means that, for the cedar, the growing season is relatively short and hardly extends beyond a single month in the High-Atlas and continental Mid-Atlas regions, beyond one to three months in the Atlantic Mid-Atlas region or beyond two to three or four months in the Rif (Mhirit 1999). Variations also depend on the altitudinal position of each individual relief.

15.4.2. *Impact on the growing season and distribution area of the cedar*

Based on estimations of climate change in Morocco (an increase of 1°C in temperature and a 4% reduction in precipitation by 2020, and an increase of 2°C–4°C by 2100), it is possible to simulate the consequent impact on the cedar's geographical and altitudinal distribution. According to this scenario, the reduction in precipitation will alter the poles between which precipitation levels oscillate for all cedar forests; these values will not exceed 528 mm–1,728 mm per year, as opposed to the current 550 mm–1,800 mm. The annual average minimum and maximum temperatures will be 4.6°C and 19.2°C instead of 3.6°C and 18.3°C respectively. These modifications will lead to a disruption of the seasonal signal, which in turn will have the particular effect of reducing the growth period (shortening or even fragmenting the growing season) and changing the spatial contour of successive vegetation zones, of the cedar.

To understand this latter aspect, the conceptual biogeographical law (hypothesis) of this model associates a change of 3°C with a change in altitude of 500 m (Tessier 1999). In our case, the retained law associates a displacement of 150 m with an increase in temperature of 1°C by 2020, and a displacement of 500 m with an increase of 3°C by 2100. The altitudinal amplitudes of zones of cedar, notably the supra-Mediterranean and the mountainous Mediterranean, which may result from these hypotheses are shown in Table 15.3.

The impact of climate change can lead to the loss of the coldest climatic zones and the linear displacement of vegetative zones towards summits. Consequently, the cedar will disappear from important areas within its current range, as evidenced by the diebacks in process in the Mid-Atlas. By 2020, supra-Mediterranean ecosystems will be reduced to a zone around 250 m in altitude, and of 350 m in the

Mediterranean mountains, in the Rif and the Mid-Atlas as much as in the more fragile High-Atlas.

Vegetation series of the cedar by zone	Supra-Mediterranean			Mediterranean mountains		
	Rif	Mid-Atlas	High-Atlas	Rif	Mid-Atlas	High-Atlas
Current altitudinal limits	1,400–1,800	1,600–2,000	1,800–2,200	1,800–2,300	2,000–2,500	2,200–2,700
Limits in 2020 (150 m/1°C)	1,550–1,800	1,750–2,000	1,950–2,200	1,950–2,300	2,150–2,500	2,350–2,700
Limits towards 2100 (500 m/3°C)	–	–	–	–	–	–
Upper limit of forests	2,400	2,600	2,800	2,400	2,600	2,800

Table 15.3. *Simulation of the altitudinal limits of cedar vegetation zones by 2020 and 2100*

As these zones contain an important part of the endemic flora of the area (Benabid and Fennane 1999), the impact on the mountains' biodiversity will be highly significant, above all because the cedar forests' floristic wealth is estimated at a thousand species, nearly 10% of which are trees, 15% are shrubs and 75% are annual or perennial herbaceous plants. The impact on the altitudinal distribution of the vegetation cannot be analyzed without taking into account its interference with latitudinal distribution (Quézel and Berbero 1990; Quézel 2001).

Particularly at low altitudes, Mediterranean species could be substituted for hillside species. On the south faces, warming and a reduction in precipitation could lead to a northward progression of Mediterranean ecosystems (ecosystem "steppization"). These phenomena are likely to have a more intense ecological impact, particularly a northward movement of the Saharan climate and its associated species (Quézel et al. 1990). The forecast for 2100 foresees the complete disappearance of the cedar from its area of origin.

15.5. Conclusion

If the displacement of the distribution areas of forest species and the alteration of altitudinal limits constitute the primary risks associated with climate change, we can be sure that forest management will be greatly disrupted. If we consider only massive tree death and diebacks, the primary consequences for forest management will be, among others:

– an increase in diebacks and the appearance of widespread deaths, with an increased risk of the proliferation of bioaggressors and of fires;

– an increase in the amount of wood to be harvested as part of clean-ups and management;

– a questioning of management and governance plans and changes to the defensive enclosures relating to the enlargement of regeneration groups, etc.;

– an insufficiency with regard to equipment and service networks, and wood clearance and emptying routes;

– an increase in the traffic of machines through stands, and a consequent increase in soil compaction and harm to remaining trees;

– an insufficiency with regard to forest management teams (human resources) available for the planning of cuts, the task of marking trees for felling or conservation and for clearing.

Other consequences can appear alongside those listed, particularly concerning the cedar industry, its commercial capacities and the economic markets for the resulting products. Conflicts and tensions between actors within the forest and administrations are at risk of multiplying, particularly those between users of forest paths, forestry cooperatives and rural communities, above all with regard to the enclosure of regenerated areas and areas marked out for new plantations, water sources and points, etc.

15.6. References

Agoumi, A. (2005). La vulnérabilité hydrique du Maroc face aux changements climatiques : la nécessité des stratégies d'adaptation. *Objectif terre. Bulletin de liaison du développement durable de l'espace francophone*, 36–38.

Aussenac, G., Granier, A., Bréda, N. (1995). Effets des modifications de la structure du couvert forestier sur le bilan hydrique, l'état hydrique des arbres et la croissance. *Revue forestiere française*, 47(1), 54–61.

Benabid, A. and Fennane, M. (1994). Connaissances sur la végétation du Maroc : phytogéographie, phytosociologie et séries de végétation. *Lazaroa*, 14, 21–97.

Benabid, A. and Fennane, M. (eds) (1999). Principales formations forestières au Maroc. In *Le grand livre de le forêt marocaine*. Mardaga, Brussels.

Breda, N. (1999). L'indice foliaire des couverts forestiers : mesure, variabilité et rôle fonctionnel. *Revue forestiere française*, 51(2), 135–150.

Breda, N., Lefèvre, Y., Badeau, V. (2002). Réservoir en eau des sols forestiers tempérés : spécificité et difficultés d'évaluation. *Houille blanche*, 3, 25–40.

Brocher, P. (1977). La sécheresse de 1976 en France : aspects climatologiques et conséquences. *Hydrological Sciences – Bulletin des sciences hydrologiques*, 22(3), 393–411.

Dajoz, R. (1996). *Précis d'écologie*. Dunod, Paris.

Déqué, M. (2000). Modélisation numérique des changements climatiques. Impacts potentiels du changement climatique en France au XXIe siècle. Mission Interministérielle de l'Effet de Serre. Report, Ministère de l'Aménagement du Territoire et de l'Environnement, 22–45.

Derrak, M., Mhirit, O., Mouflih, B., Et-Tobi, M. (2008). Influence de la densité et du type de peuplement sur le dépérissement du cèdre de l'Atlas à Sidi Mguild (Moyen Atlas marocain). *Revue forêt méditerranéenne*, 29, 1.

Dreyer, E., Aussenac, G., Granier, A., Guehl, J.M. (1992). Sécheresse et physiologie des arbres. Les recherches en France sur les écosystèmes forestiers. Report, Ministère de l'Agriculture et de la Forêt, Paris.

Et-Tobi, M. (2006). Approche multidimensionnelle des relations "état sanitaire-station – sylviculture" pour l'étude du dépérissement des cédraies (CedrusatlanticaMan.) au Moyen Atlas en vue d'élaborer un modèle sylvicole de prévention phytosanitaire. PhD Thesis, IAV Hassan II, Rabat.

Et-Tobi, M. (2008). Inventaire dendrométrique et phytosanitaire des cédraies d'Azrou et Ait Youssi Lamekla. Étude des causes du dépérissement de la cédraie du Moyen Atlas (Ifrane). Report, FAO, UTF, MOR, HCEFLCD.

Et-Tobi, M. and Mhirit, O. (2003). Productivités forestières et modèles de croissance du cèdre de l'Atlas. PhD Thesis, IAV Hassan II, Rabat.

Et-Tobi, M., Mhirit, O., Mhamdi, A. (eds) (2006). Concepts, définition et prédictions des dépérissements forestiers. In *Le cèdre de l'Atlas. Mémoire du temps*. Mardaga, Brussels.

Gibelin, A.L. and Déqué, M. (2003). Anthropogenic climate change over the Mediterranean region simulated by a global variable resolution model. *Climate Dynamics*, 20, 327–339.

GIEC (1997). Incidences de l'évolution du climat dans les régions : évaluation de la vulnérabilité. Résumé à l'intention des décideurs. Report, Groupe de travail II du GIEC, UNEP, PNUE, WMO, OMM.

GIEC (2000). L'utilisation des terres, le changement d'affectation des terres et la foresterie. Résumé à l'intention des décideurs. Report, Groupe d'experts intergouvernemental sur l'évolution du climat, PNUE, OMM.

GIEC (2001). Bilan 2001 des changements climatiques : conséquences, adaptation et vulnérabilité. Report, Contribution du Groupe de travail II au troisième rapport d'évaluation du rapport, Groupe d'experts intergouvernemental sur l'évolution du climat, PNUE, OMM.

GIEC (2007a). Bilan 2007 des changements climatiques : impacts, adaptation et vulnérabilité. Report, Contribution du Groupe de travail II au quatrième rapport d'évaluation du Groupe d'experts Intergouvernemental sur l'Évolution du Climat. Résumé l'intention des décideurs.

GIEC (2007b). Bilan 2007 des changements climatiques : les bases scientifiques physiques. Report, Contribution du Groupe de travail I au quatrième rapport d'évaluation du Groupe d'experts Intergouvernemental sur l'Evolution du Climat. Ministère des Affaires Etrangères, Ministère de l'Ecologie et du Développement Durable, Ministère Délégué à l'Enseignement Supérieure et à la Recherche.

Granier, A., Badeau, V., Bréda, N. (1995). Modélisation du bilan hydrique des peuplements forestiers. *Revue forestiere française*, 57, 59–68.

Granier, A., Bréda, N., Biron, P., Villette, S. (1999). A lumpted water balance model to evaluate duration and intensity of drought constraints in forest stands. *Ecological Modelling*, 116, 269–283.

HCEFLCD (2006a). Description et analyse des conséquences du phénomène de dépérissement. Étude du dépérissement du cèdre. Projet d'aménagement concerté des forêts et parcours collectifs de la province d'Ifrane. Report, HCEFLCD, DREF-MA.

HCEFLCD (2006b). Stratégies et programmes d'intervention des cédraies dépérissantes. Étude du dépérissement du cèdre. Projet d'aménagement concerté des forêts et parcours collectifs de la province d'Ifrane. Report, HCEFLCD, DREF-MA.

HCEFLCD (2006c). Guide de sylviculture des cédraies dépérissantes. Étude du dépérissement du cèdre. Projet d'aménagement concerté des forêts et parcours collectifs de la province d'Ifrane. Report, HCEFLCD, DREF-MA.

Hoff, C. and Rambal, S. (1999). Les écosystèmes forestiers méditerranéens face aux changements climatiques. *Comptes rendus des séances de l'Académie d'agriculture de France*, 85(4), 53–57.

Houghton, J.T., Meira-Filho, L.G., Callander, B.A., Harris, N., Kattenberg, A., Maskell, K. (1996). *Climate Change: The Science of Climate Change*. Cambridge University Press, New York.

Landmann, G., Bréda, N., Houllier, F., Dreyer, E., Flot, J.L. (2003). Sécheresse et canicule de l'été 2003 : quelles conséquences pour les forêts françaises ? *Revue forestiere française*, 55, 299–308.

Mateuh (2000). Le territoire marocain : état des lieux. Report, Contribution au débat national sur l'aménagement du Territoire, Ministère de l'Aménagement du Territoire, de l'Environnement, de l'Urbanisme et de l'Habitat, Mateuh, Rabat.

Médail, F. and Quézel, P. (1996). Signification climatique et phytoécologique de la redécouverte en France méridionale de Chamaerops humilis L. *Comptes rendus de l'Académie des Sciences de Paris*, 319, 139–145.

Mhirit, O. (1982). Étude écologique et forestière des cédraies du Rif marocain. Essai sur une approche multidimentionnelle de la phyto-écologie et de la productivité du cèdre. PhD Thesis, Aix-Marseille Université, Marseille.

Mhirit, O. (1994). Croissance et productivité du cèdre de l'Atlas : approche multidmensionnelle de l'étude des liaisons stations – Productions. *Annales de la Recherche Forestière au Maroc*, 27(1), 296–312.

Mhirit, O. (ed.) (1999). Climats et bioclimats de la forêt. In *Le grand livre de la forêt marocaine*. Mardaga, Brussels.

Mhirit, O. and Benchekroun, F. (2006). Les écosystèmes forestiers et périforestiers : situation, enjeux et perspectives pour 2025. Report, Contribution au rapport sur le Développement Humain.

Moreno, J.M. and Oechel, W.C. (eds) (1995). *Global Change and Mediterranean-Type Ecosystems*. Springer, Belin.

ONF (1999). L'Eau et la forêt. *Bulletin technique de l'ONF*, 37, 240.

Quézel, P. (2001). Biodiversité végétale des forêts méditerranéennes, son évolution éventuelle d'ici à trente ans. *Forêt méditerranéenne*, 2–8.

Quézel, P. and Barbero, M. (1990). Les forêts méditerranéennes : problèmes posés par leur signification historique, écologique et leur conservation. *Acta Botanica Malacitana*, 15, 145–178.

Quézel, P., Barbero, M., Bonin, G., Loisel, R. (eds) (1990). Recent plant invasions, in the Circum-Mediterranean region. In *Biological Invasions in Europe and Mediterranean Basin*. Kluwer Academic Publishers, Dordrecht.

Rambal, S. and Debussche, M. (1995). *Water Balance of Mediterranean Ecosystems Under a Changing Climate. Anticipated Effects of a Changing Global Environment on Mediterranean-type Ecosystems*. Springer, Berlin.

Tessier, L. (1999). Impact des changements climatiques en montagne. Report, Laboratoire de Botanique historique et de Palynologie, ERS CNRS 6100, Agora21 ARMINES/ENSM-SE.

16

The Legacy and Future of Conservation in El Kala National Park (Algeria)

16.1. Introduction

Nature conservation is developed in situ based on the practical conditions which determine human uses of an ecosystem, and also through the historic and institutional trajectories that shape a given area's conservation apparatuses. Today, conservation is seen as polysemic in nature, a polysemy linked to conceptual and practical changes in its implementation. If conservation has previously consisted of the delimitation of natural spaces for the purpose of protecting them from human activity, it now covers a range of rules for the use of nature that ought to be tied to the ways of life of the populations living within or on the peripheries of protected areas. Since the Convention on Biological Diversity (1992), protected areas have been opened up to usages considered compatible with nature conservation, and article 8J of the Convention associates the conservation of nature with that of culture.

The link between these two dimensions of conservation, while also introducing ambiguities (Dahou and Wedoud Ould Cheikh 2007), has manifested the idea that those populating protected areas are the most capable of preserving the nature therein, through uses and knowledge developed in close relation with the environment. Nevertheless, these provisions by international conservation regimes are applied in a particularly different manner depending on the local territory in which they are promoted, not only because of local social and ecological peculiarities, but also because of particular historical trajectories.

Chapter written by Tarik DAHOU.

UNESCO's Man and the Biosphere Program can be seen as a precursor in this regard, not only in its attempt to marry conservation and sustainable development, but also in its search to further integrate different kinds of knowledge on biodiversity, both scientific and local. This approach can be seen in a great number of biosphere reserves, but with very different results depending on the situation and on the difficulties encountered in organizing solid collaborations between managers and populations (Dahou et al. 2004). In Algeria, the El Kala National Park (EKNP), created in 1983 in one of the largest humid zones in the Mediterranean, was designated as a biosphere reserve by UNESCO in 1990, with the objective of managing interactions between nature and society. Beginning in 2005, a project for the creation of a Marine Protected Area was initiated along its coastal areas, with the elaboration of a management plan under the aegis of the park's authorities.

While Marine Protected Areas are essentially a development of the new millennium, they have the same characteristics as those on land. Conservation models developed on dry land have been transposed onto the sea, notably with regard to zoning. In the particular case of the EKNP, the Marine Protected Area project was borne by the park institution itself. This situation, visible in other Marine Protected Areas in which the park model is sometimes replicated, is not systematic. Nevertheless, it does put into question the long trajectories of these park-style nature conservation models, as much on an international scale as on Algerian territory. While the Marine Protected Area model is not often characterized by an adaptation of their management rules to local ecological and socio-economic contexts (Jentoft et al. 2007), its institutional trajectories are no less influenced by a historicity (Walley 2004) that it is a question of recovering.

The case of the EKNP is interesting, as it opens onto a history of conservation in Algeria, replicated in part in the French colonial empire. It is also marked by the entrenchment of park management rules in the strategies of the capitalist colonial model. We will first question the trajectories of conservation in Algeria and then those of the park institution, so as to discern the traces they have left on the management of the EKNP and their influence on how it imagines the management of its marine area.

We will thus attempt to identify how the trajectories of conservation and sustainable development policies depend on different legacies expressed within the different parts of the state, which lay out contradictory modes of management within the area and prevent the conciliation of knowledge.

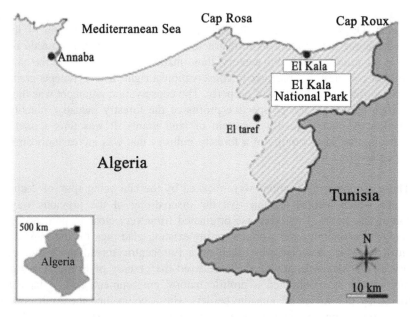

Figure 16.1. *Location of the El Kala National Park*

16.2. Declinism, forest exploitation and management in the EKNP

Conservation has been developed precociously in Algeria, as it has allowed for the redistribution of usage rights and ownership so as to construct military and political colonial power. Immediately after the conquest of Algeria, means were produced to prevent indigenous use of agricultural and forest spaces. They responded to different logics, united by a conservationist discourse inspired by declinism, which had as its primary objective the justification of dispossession, as Diana K. Davis (2007) has clearly shown.

After the conquest, the colonial state's expropriation strategies led to the reduction of the land structure to a *tabula rasa*, as users without written rights were evicted from their lands. This is notably the case for ownership or usage rights for collective tribal lands or collective lands dedicated to extensive breeding, particularly forest spaces. This process began with a recognition of private lands, *melk*, having benefited from a statute of propriety under the Ottoman yoke, or religious properties, *habous*, lands governed on pious grounds. Later, indigenous rights would be recognized for tribal lands within the framework of the 1863 *sénatus-consulte* under Napoleon III, but this law would in fact pave the way for the process of delimitation and transfer to state or private ownership that would be accelerated from there (Guignard 2010).

The appropriation of natural resources first served exclusively political objectives (Davis 2007, p. 32). Military efforts made the exploitation of forests unavoidable for the creation of infrastructures that would meet the needs of the army. In the middle of the 19th century, these maneuvers were made in the name of the right of conquest, but recourse to conservationist rights and arguments gradually legitimized this reorganization of property. The conservation argument was thus put forward to perpetuate the prospective profits of the forestry industry, notably the wood industry, and to justify a system of land grants. It was thus a matter of guaranteeing the sustainability of a forestry industry that was increasingly strategic for the colony.

These economic motivations were masked by the tale being spun of decline, a discourse which had as its objective the discrediting of the previous ways of exploiting the forest. This discourse originated from very close to political circles and colonial economics and would swap its existing glad rags for the trappings of scientificity. This discourse was based on a Eurocentric rereading of the natural history of the Maghreb, previously considered the granary of Rome. The Roman Empire would have inherited a prolific natural environment said to have been destroyed by the hordes of nomadic herders who accompanied the Arab invasion. By reappropriating the image of a Hilalian catastrophe first depicted by Ibn Khaldun[1], the French saw only destruction in the Arab conquest of the Maghreb. They willingly assimilated the idea of the conquest with that of a rupture from the ancient agropastoral model that sat in equilibrium with nature, a rupture in favor of an extensive breeding model, considered to have been destructive of the forest.

Behind this myth of the decline of nature in the Maghreb, a decline that took the form of an ineluctable desertification hid the specter of the eradication of indigenous usage rights. The myth declared that deforestation was unrelated to the climate, but was attributable to overgrazing and to the clearings by fire aimed at the fertilization of rangelands. Only conservationist measures and plantation programs could recover the grandeur of nature and indigenous property rights were thus expelled from forest lands, including lands included in ancient forests.

The colony thus presented itself as a protector of nature, while conferring on itself alone the rights to the land in the name of a genealogical link with the Roman Empire. This myth evidently obscured colonial deforestations carried out for military and economic purposes. It served above all to lend legitimacy to the

---

1 The conquest of the Maghreb by the Arabs (the Banou Hilal tribe) in the 11th century was thus described as a catastrophe, leading to the devastation of institutions and of the economy, thanks to the incompatibility of the nomadic way of life with the agricultural and urban civilizations that characterized the Maghreb. Numerous sources temper, or even invalidate, this representation of the Hilalian invasion.

expropriations at the base of the constitution of a national domain open to new modes of exploitation. Scientists were summoned to endorse this discourse. Natural science confirmed this image of a forest decline, denying, in large part, the particular nature of the Maghreb climate and the resistance of ecosystems to risks, including fires, grazing and clearings (Davis 2007, p. 12). It contributed to the formulation of a certain number of conservation and plantation measures within Algeria.

In this framework, forest conservation was justified as much by the restoration of the fertility of the land, to deal with erosion, as by the control of flooding or even by hygienist arguments (Ford 2008). The new vocation of the forested rangelands of the indigenous peoples was to help manage water on the plains held by colonizers and to help maintain the health and well-being of the occupants. The management of the environment was thus closely tied to that of hydraulics, particularly after the great drought in the middle of the 19th century, which not only led to a severe famine, but also to a large-scale cholera epidemic (Cutler 2014). The restoration of forest spaces also led to a plantation policy aimed at the draining of the marshes and humid zones that were considered improper for the populating of cities and to economic industry. These environments were considered to be vectors of diseases deadly to the colonizers, their economic importance being realized only later.

These different points of view were deployed through the intermediary of colonial sciences, and led to conservation and plantation projects regarding forest zones and zones destined for forest regeneration. The ancient rights holders were evicted from forest zones and penalized with convictions and fines if they infringed cantonment rules or made recourse to fires to try to restore their pastures. While the forest code criminalized the use of the forest, it perpetuated states of exception. Numerous forests had been for a long time managed by the military institution (for martial reasons relating to construction and to territorial control), and the reconstitution of massifs was carried out under the aegis of military planters (Davis 2007, p. 79). This Algerian trajectory was at the origin of the forest conservation that we can find throughout the French colonial empire.

However, this history of conservation in Algeria is closely associated with the history of the silvicultural industry's exploitation of the forest, in alignment with preoccupations with the densification of economic and political infrastructure as well as for developing the exportation of its products to mainland France. We can see here the convergence of public and private interests after the conquest, which is also the case for the financing and profitability of hydraulic infrastructures (Cutler 2014). This industry relied on a regime of land grants or on the privatization of forests with the standard of reintroducing species, this last option being imposed through colonial planting. The industry was comprised not only of the exportation of wood, but also of sub-products, such as cork oak bark, which were at the heart of silvicultural commerce between Algeria and mainland France. Colonial silvicultural lobbies were also preoccupied with plantations, as they were comprised of the same people who were championing declinist rhetoric, as well as the capacity for plantations to restore a favorable environment for lucrative enterprise. This led to the introduction of the eucalyptus tree to Algeria, the species being planted to dry out soils in humid zones (Davos 2007, p. 61). While there were voices of dissent, particularly among botanists (Davis 2007), they were soon converted by military and economic pressure groups.

The field of conservation is thus closely correlated with the militarization of Algeria and the forestry industry. Furthermore, it was the forest zones administered for silvicultural needs that were destined to become parks within colonized Algeria. The institution of the park in Algeria emerged from this context, marked by a will to separate the indigenous peoples from nature, to the profit of colonial society. French parks were first created in the colonies in which political restraints around land ownership were less acute (Selmi 2009), the rights holders having been suppressed by force. A similar imperial logic of conservation manifested in this period in the British empire, particularly in India (Agrawal 2005), to perpetuate the exportation of wood to mainland Britain.

Contrary to American models aimed at maintaining a wild nature – the wilderness (Harper and White 2012; Jones 2012) – forestry occupied an important role in park management. In Algeria, local uses tended to be excluded from forest zones that had entered state ownership, while certain uses were sometimes tolerated, at least until the application of the French Forestry Code in 1883 (Guignard 2010). The hygienist dimension of colonialist policy, which took shape at the end of the 19th century in the colonies, placed a strong emphasis on the importance of the forest in containing desertification and indigenous insalubrity, instead favoring the growth of the Euro-Mediterranean population – a preoccupation more important here than in the rest of the French empire, revealing the dark side of colonial environmentalism (Ford 2008).

Despite the power of the silvicultural lobby, alternative forms of forest management were envisaged at the start of the 20th century. Forest reserves, created to scientific ends, were also in the picture from 1913, thanks to the universalist vocation of French science and its representatives in Algeria. A different model, however, became predominant: that of the park, which associated conservation with aesthetic valuation. A series of parks were founded in Algeria from 1921 onwards, the management perspectives of which were concerned above all with frequentation by tourists (de Peyerimhoff et al. 1937). Park creation was thus a matter of promoting the colonial project, assimilation and territorial unification (Zytnicki 2013). Few scientific reserves would go on to form parks, not even the 10 hectares envisaged at Calle (today El Kala (Aubreville et al. 1937)), to the extent that the preservation of spaces for scientific practice did not figure into the priorities of administrators.

16.2.1. *The legacy of declinism in the EKNP*

Immediately after the conquest, colonial documents began to show an interest in the cork oak forests of Calle[2], emphasizing low agricultural land occupation (arboricultural and regarding peanut growth), enclosed into small areas in the footlands[3]. Exploration allowed for the analysis of land developments with high settlement potential. They also noted the possible exploitation of this area through inland fishing. While no park was created in El Kala in the wake of the 1921 law, forest conservation measures appeared during the colonial period (PNEK 2011), as they did in other areas.

The space now known as the El Kala National Park (EKNP) bears all the scars of colonial nature conservation policies. The different modes of exploitation once deployed on this national land forged its nature and forest landscapes. The cork oak was greatly exploited in El Kala, given its endemicity to the mountainous massifs therein. But along the lakeshores, we can also see eucalyptus trees were planted to dry out humid lands. The El Kala region is a herding zone once occupied by tribal groups who had lived in the region for a very long time. Colonialization had extended its state-owned land here above all to the detriment of pastures. The organization of the land was thus no longer dependent on the tribe, desubstantialized by colonial land reforms (Hachemaoui 2013), particularly in coastal zones; in the steppes, the collective territories were less subject to colonial land seizure (Ben Hounet 2013). With the protection of forest zones, enclosure against indigenous use was established, in a region previously characterized by extensive breeding and by a close articulation of routes between forest zones and lakeshores. Colonial forest exploitation was, however, disturbed in this region by fires, to the point where former concessionaries returned their forests to state ownership (Tomas 1969). These fires may have been attributable to herders wishing to continue to use their ranges.

Local ranges were also affected by the expansion of agricultural exploitation in the wilaya (the equivalent of the prefecture), which grew from 6,000 to 14,000 hectares from the middle of the 19th century to the middle of the 20th century (Homewood 1993). Lake Tonga was drained to increase available farmland,

---

2 Duvérine, A. (1840). *De la gestion des intérêts français en Afrique. Résumé critique de l'état politique et économique de l'Algérie.* Ledoyen, Paris.

3 Bérard, V. (1867). *Indicateur général de l'Algérie, description géographique, historique et statistique de toutes les localités comprises dans les trois provinces.* Bastide, Alger, p. 383.

primarily dedicated to growing peanuts, tobacco, cotton, and, to a lesser degree, to market gardening. Despite the restrictions on pastures introduced by forestry operations, livestock practically doubled in the century of colonization, accompanying population growth.

The independence period renewed these models of nature use, rather than questioning them. In place of sharecropping, cooperative agricultural industry was introduced, and clearings were carried out so as to increase available farmland. The cork industry was also renewed, while extensive breeding continued to be ignored; forest legislation was not amended, perpetuating the theoretical incompatibility of herding practices with maintenance of the forest. Despite the importance of livestock rearing in household economics, no proposals were made for the management of herding spaces. On the contrary, forest zones were the object of particular attention with regard to reforestation and to the creation of firebreaks and tracks for cork harvesting campaigns. Foresters also busied themselves with the planting of fast-growing species intended to supply a future paper pulp factory (Tomas 1969). Despite the changes in land use between the colonial and independent periods, livestock rearing continued to be carried out within the region.

The EKNP was created in 1983 and covers 76,000 hectares, 10% of which are bodies of water, within the large humid zone of the El Tarf wilaya (near the old El Kala district). It is inhabited by 77,000 people, residing in several urban communities and douars (groups of homes and base administrative divisions) around pastoral and agricultural lands, and roughly encompasses the El Kala, Souarekh, El Aïoun, Ramel Souk, Aïn Assel and Bougous municipalities, in which a little over 50% of the population is urban.

For example, agricultural lands are situated in areas managed by the water authorities, which have changed from buffer zones to protected areas through successive management plans, because of their contiguity with bodies of water, which are integral protected areas. These ambiguities of zoning are not clarified by particular apparatuses – cultural norms to be respected – in the management plan.

As the park does not own the land it covers, its creation instead ratified a prior occupation of the space and previous legal divisions, notably while placing protected areas, made up, on the one hand, of forest zones coming from the forestry service, and, on the other hand, of the Oubeira and Tonga lakes, and the Mellah lagoon, coming from the water authorities. Its role is in fact limited to coordinating the actions of other services, those services being the owners of the land. It remains confined to carrying out conservation activities in the remaining spaces. In 1987, 73% of the park's land was state-owned forest land managed by the forestry service, while 9% was state-owned agricultural land, 10% was private land and 7% belonged to the water authorities (Homewood 1993). The land ownership structure of the park rests unchanged (PNEK 2011) and the management plan applies differentiated protection statuses.

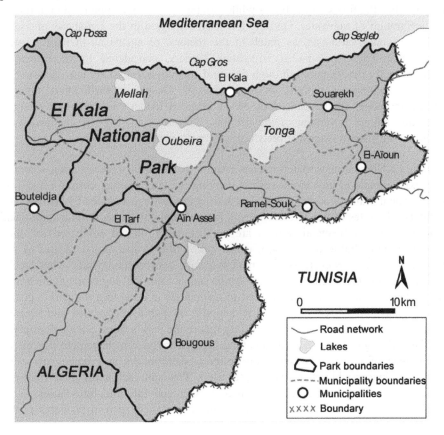

Figure 16.2. *El Kala National Park and its municipalities*

The park's bodies of water are part of the integral protected area. Dune areas benefit from a "wild area" status, which theoretically limits activities in coastal areas. Forest zones are equally considered conservation zones, industry usually being limited only to silviculture. We can see here a division modeled on the spatial distribution of ecosystems – aquatic, dune or forested. While the park's agricultural lands are situated in protected areas, it is difficult to see which spaces are dedicated to livestock, which still works on an extensive breeding model deployed between the shores of bodies of water and contiguous forest spaces.

Despite the protection statuses that have limited mobility, we can identify a relative stability with regard to numbers of heads since the colonial period

(Homewood 1993). There is a real complementarity between herding and agriculture. In the winter, the animals graze in the forests, and then, from spring, they fertilize the fields on the banks while grazing on the vegetation built up through the infiltration of rainwater. Herding has itself remained in the crosshairs of forest and park conservationists, as much in the rhetoric of agents as in management documents. These documents stigmatize herders, considered to be primarily responsible for deforestation through fires and grazing, never taking into account the impact of climatic variations on forest massifs. Despite intensification projects around the edges of the park, focused on irrigated fodder cultivation, originating from an umpteenth version of the declinist discourse, nothing has changed with regard to the ancient extensive breeding system (Homewood 1993).

The EKNP continues to define conservation on the basis of area-sharing schemas that exist outside of any social reality, denying the land's actual current uses. The management plan advocates a model of zoning entirely preoccupied by species and habitats. The rules of the park are mostly ignored, not only because its institutional, land-owning and financial powers are comparatively limited compared to other administrations, but also because it makes little effort to accommodate uses of the land which are, however, even compatible with conservation goals. This feeble integration of the logic of nature conservation and the logic of maintaining non-resource heavy local uses has led to difficulties in accepting the park's goals. These goals have been only weakly adopted by public authorities and elected officials, while residents seem to willfully ignore the park's rules. Such a restriction on the park's goals with regard to natural or even touristic management is a direct inheritance of colonial conservation policies, with forests and bodies of water remaining inaccessible in theory to local users. This approach is divorced from the reality of these uses of nature, which remain crucial for rural households, as is shown by the study of the different types of development seen in residents of the park.

16.2.2. *Uses of the EKNP's natural resources*

An analysis of the use of natural resources in the park was carried out in 2009, beginning with extensive investigations in the coastal park of the park, based on individual semi-directed interviews carried out in rural areas within the EKNP, covering the El Kala, Souarekh, Ramel Souk and Aïn Assel municipalities. This study went on to form the basis on which a quantitative investigation was developed, being carried out in 2010[4]. At the end of the 1980s, 40% of the income of El Kala's

---

4 The surveyors dealt with a sample of 600 rural households from douars in the relevant municipalities. The households were chosen at random by questioning 10% of the households in each area of rural residence (douars or small rural communities), without extending this

population, including the urban population, came from agriculture (De Belair 1990). This number is possibly a little lower now, as employment has diversified in the agricultural zone, which has seen intensified littoralization (PNEK 2011), but agriculture has also been able to profit from the liberalization of the economy.

16.2.2.1. Agriculture

The cultivated agricultural areas of the EKNP are not large in scale thanks to restrictive zoning, and abundant winter rains soaking the lowlands of this humid zone for half of the year. On many plots, agriculture is limited to the summer months. Analysis of our survey sample shows that around half of rural households practice agriculture (peanuts, melons and watermelons).

Each year, few areas are developed, as on average they only reach around a hectare in size per household, insofar as these lands, belonging to households, are reduced, and there are no structures for financing them. Around half of cultivated lands are privately owned, and a quarter are lands issued from the state's private domain, given as land grants through individual farm statuses. The majority of farmers use water from the lake closest to them, or from wadis near to their fields, which tends to limit dewatering costs (only 10% of farmers make recourse to wells). To this relative compression of agricultural areas is added weak intensification, phytosanitary products rarely, or only in small doses, entering into the technical plan. Agricultural yields have proven to be feeble, with an average of a metric ton per hectare across all crops.

The low productivity of cultural practices and the narrowness of plots (low water resource consumption) limit their impact on the environment. Pollution is negligible, given the sparse usage of chemical products and the use of natural fertilization by herds. Manpower is mostly supplied by members of the family. Agriculture contributes to household reproduction to varying degrees, but for those who practice it, the revenues from agriculture still constitute a non-negligible part of the household's earnings.

16.2.2.2. Herding

This agricultural practice is combined with the practice of extensive breeding of bovines, sheep or goats, even though herding and agriculture are not always

---

work to the main urban centers of these municipalities. As the four municipalities chosen are inhabited, following the most recent census in 2008, by 57,000 people, and as the urbanization level of these municipalities is 52%, the rural people in these municipalities come to approximately 27,000 people. As the average household size hovered at six members, we can estimate that our sample covered more than 10% of the rural households in these municipalities.

integrated. Two thirds of our sample practice herding; numerous households thus possess heads of cattle without practicing agriculture and put their animals to pasture on their uncultivated agricultural land. More generally, the animals frequent pastures situated on the state-owned lands in the forested areas of the park. They can be found there for a good part of the year, outside the dry period, in which the cattle congregate along the shores of the lakes before the lands there are cultivated. The size of the herds is in keeping with the rarity of pastures in El Kala's humid zone, with three quarters of households having fewer than 20 heads (of all species together).

While some of this production is sold at local markets by the herders, only a small part of the livestock is sold throughout the year. The income garnered through selling cattle is meager and stands as extra income. Non-intensive herding is above all motivated by economy in a region in which production costs are low thanks to the fodder provided by forests and the water resources abundant in the humid zone.

16.2.3. *The structure of rural revenue in the EKNP*

In the EKNP's rural zones, around half of all household revenue comes from the exploitation of natural resources through agriculture or herding. Inhabitants garner no revenue, however, from exploitation of the forest, which falls under concessions granted by forest conservation to business owners. Neither do they take forest products for their own consumption in addition to pastoral resources. Another revenue source is connected to fishing, as some inhabitants catch a small number of fish in the park's bodies of water or are even engaged in marine professions in El Kala.

These statements contradict the preparatory documents for the different management plans for the EKNP's land and marine areas, which emphasize the small impact of human activity on nature. Even though the impacts of these activities are far from endangering the renewal of natural resources in the humid zone, the ecosystems of the park are strongly anthropized. Nevertheless, this revenue structure indicates that it would be difficult for rural households to ensure their continuation based solely on productive activities linked to nature. Even though around half of all revenue comes from such activities, they prove feeble and rely on strong diversification within the household.

Inhabitants generate a range of revenues by combining the different activities of different members of the household. Many households benefit from salary revenues in a region where scarcity of employment has pushed inhabitants to join the military or the police. There are many young people who have retired from the army and benefit from pensions. Formal employment can be found in the fishing industry, in the port of El Kala or in transport. Informal employment is more prevalent, however, in the construction sector or in security. Rural households' standard of living, which is relatively low, leads them to diversify their revenue sources by multiplying activities, but the exploitation of natural resources remains a determinant factor in their reproduction.

16.2.4. *Conservation for improved exploitation*

Rural populations' uses of nature thus appear to be highly diversified, between an agropastoralism and complementary revenues found in inland and maritime fishing, or even more marginally by forest products. Beyond recourse to salaried activities, households have the capacity to diversify their development of natural resources. This latter option relies on the deployment of the different members of the household to different types of exploitation which are integrated together through the intermediary of the workforce and of self-financing, given the rarity of credit. Households adapt to the constraints of the ecosystems (seasonal rhythms, climatic environment and availability of resources) through the multi-localization of activities, breaking from the compartmentalization of the use of resources which underlies the zoning and the management plan of the EKNP.

These uses reveal a certain paradox between a planning-based approach, which accords little importance to managing the spatial and temporal continuity of interactions between natural and social dynamics, and the reality of the practices of those living in the park's rural spaces. The modes of agricultural and pastoral exploitation are still based on familial industry, always reliant on extensive breeding practices and always with recourse to salaried revenues from formal or informal work. The close imbrication of these different types of revenue is far from suggestive of a transition towards nonagricultural activities. Rather, it speaks to the importance of the exploitation of nature to local ways of life, as it is in large part financed by salaries and transfers (from migrants living elsewhere in Algeria or from pensions).

Land occupation and uses reveal the discrepancy between the park's territorial structure, derived from naturalist principles, and the practices of the population, which spills over and out from the normative spatial organization of the park, particularly onto its state-owned spaces. They elude to the imposition of a virile, quasi-military planning of nature, a typically French approach[5] (Larrère et al. 2009),

---

5 We can always see a strong trend towards the exclusion of ancient uses in the creation of parks, organized around a redefinition of inhabitants' uses by force, whether in an American context, based on the idea of the wilderness, or in contexts more marked by modern European history (West et al. 2006).

forged into the Algerian colony through the imposition of a naturalist order. The act of creating a park indeed stems from an establishment of national sovereignty over a space that defines the relationship between society and nature (Blanc 2015).

This national sovereignty was imposed on a border region, parts of which, once considered to be Tunisian territory, were later reabsorbed into Algeria. This authoritarian act, reinforced by a history of states of exception based on Algerian forests, can never completely hinder the local reformulation of interactions between actors and resources. Local practices cross the authoritarian borders of the park's zoning by taking advantage of the interstices in planning and the contradictions arising from the differing interests of deconcentrated administrations (the multiplicity of interventions by Water and Forests, agricultural services, water services and the park).

In the space of the EKNP, nature has been shaped by humanity in the context of ancient uses of the land. If the ideology of conservation has historically relied on naturalist principles which endorsed the elimination of herding (under the aegis of forest conservation), forest management in Algeria today is distancing itself from an exclusively naturalist self-understanding. However, the contemporary park model (despite it being placed under the auspices of the same minister as forest conservation, the park is under the authority of a specific directorate) is itself profoundly marked by a maximalist understanding. It is managed with a disdain for the idea of updating its apparatuses in light of the contemporary conservationist canon, which now treats actions carried out in industrial areas of parks and their core areas as a continuum (Mathevet et al. 2010).

A notion of conservation based on strict naturalist principles, manifested in the zoning and management plan of the EKNP, does in fact mask the persistence of uses which struggle to integrate conservationist dimensions or even contradict them. Localized uses connected to pastoralism, accommodating various conservation rules, are characterized by strong stability through time and do not appear to have had a particularly pronounced effect on the structure of the area. The incompatibility of the residents' uses of the land with the equilibrium of the natural environment is far from proven, given the low intensity of agricultural and pastoral exploitation. The discourse around their effects, often using rhetoric of irreparable damage to ecosystems, is not scientifically supported in the different official reports from the EKNP that seek to incriminate such activities. On the contrary, there is no mention in these reports of the development of natural spaces by the businesses that have received land concessions and exploit aquatic and forest spaces, which are far more vulnerable.

In such an approach to conservation, local knowledge hardly contributes to understandings of the relationship between humans and their environment, nor to the conception of management measures, in perfect contradiction with the experiments underway in many protected areas. Contemporary reflections on the interactions between herding, fire and biodiversity (Larrère et al. 2009) are completely ignored, and local uses are still being incriminated, even though seldom punished.

Conservationist discourse still has a role in maintaining privileges for exploiting Algeria's nature. We can therefore see that the assimilation of conservation to an underlying project – the mastery and exploitation of nature through capitalist enterprise – tends to be reproduced throughout the contemporary historical cycle, revealing the deep ambiguity of global conservation projects stemming from states. While the Rio Convention, through article 8J, promotes apparatuses for recognizing the rights of the populations living in protected areas over nature so as to avoid collusion between market actors and states to develop nature, we must state that two decades later, the conservation tools inspired by international conventions hardly guarantee equitable access to natural resources.

16.3. The spread of fishing and marine conservation in the EKNP

Today, the project of extending the EKNP to cover its marine area, which is promoting a protectionist discourse through its mimetic replication of land-based zoning principles, seems to stem from this tradition of colonial conservation, with the elaboration of a management plan which disregards uses of the marine area. While declinist rhetoric has flourished in the implementation of conservation, within forest management institutions in the past and in parks today, is it more diffused with regard to marine conservation? Can we identify the same relationship between conservation of nature and the promotion of market exploitation of natural resources with little case for their equitable distribution among those living in the park?

The question of declinism could arise in fisheries, particularly in favor of conservation programs, whether in relation to responsible fishing standards[6] or especially in relation to the implementation of marine protected areas. How then are global discourses on conservation interpreted in a case study where fishing has never reached an intensity liable to lead to overfishing? An analysis of the trajectory of the fishing industry in Algeria allows us to appreciate how the problems of marine conservation are understood in our area of study within the EKNP, and to examine the environmental discourses promoted by public powers. This approach permits us

---

6 The FAO Code of Conduct for Responsible Fishing is a code of good conduct for the rational management and conservation of fish resources, aimed at implementation in the fishing policies and legislations of UN member states.

to explore the continuity of the ambiguous links between market appropriation and the conservation of resources.

Fishing has never attained a significant intensity in Algeria, neither in economic terms, nor in terms of employment. It is hindered by structural environmental constraints. While Atlantic currents, usually gauges of significant trophic[7] levels, are active in the area, Algeria's geography penalizes fishing production. The Atlas foothills, situated on the Algerian coast, offer few natural shelters from which to embark on fishing expeditions and considerably reduce the extent of the continental shelf. Two thirds of the continental shelf, with an area of approximately 14,000 km^2, have non-trawlable floors[8]. After the coastal raids by pirates, the violence of which drove coastal populations inland, the conquest of Algeria contributed to revival of fishing activities. The colony at first tolerated Spanish and Italian fishing campaigns, and then encouraged their gradual installation along the Algerian coast, to respond to the food requirements of the colony. Until 1930, there were only around 20 French fishermen to 4,750 Spaniards, Italians, and naturalized members of these communities, and 420 Arabs (Simonnet 1961). These fishermen spread the artisanal techniques of "petit métier"[9] fishing for white fish[10] and of sardine fishing for oily fish[11].

Nevertheless, for a significant maritime front – the coast of which is estimated to stretch along to around 1,600 km – fishing production has always been negligible. Throughout the entirety of the 20th century, despite these encouragements, total production remained static at around 20,000 metric tons (Furnestin 1961). It was limited by the small number of trawlers for white fish, less than 200, and the rudimentary methods used to capture oily fish (purse seines[12] have not yet been

---

7 Relating to the nutrition of species.

8 Trawling involves the dragging, by a powerful boat, of a large sock-like net along grainy or sandy floors to take in any species moving in the area of that floor. It has proven to be rather unselective, capturing a large variety of species. Trawling requires significant mechanical power to cover long distances and take in significant quantities of fish.

9 In the Mediterranean, "petit métier" refers to coastal fishing, carried out in barely mechanized small boats. It typically targets bottom-dwelling species, either with nets or lines.

10 This refers to deep-sea demersal fish found on continental shelves close to the rocks, and which have significant market value: groupers, sea bream, porgies, bass, red mullets, forkbeards, red sea bream, etc.

11 This refers to pelagic, open-water fish that move without being tied to any particular habitat (rocks, sandy bottoms, substrates) throughout the water column.

12 Purse seines are large nets designed to encircle shoals of fish and, once thrown into the water, close at the bottom to trap fish in the water column. It then gets narrower and narrower as it is pulled back onto the boat. This requires many hands and a certain level of mechanization.

introduced). The fleet as a whole was not large, neither in terms of size nor in motor power, which, despite some progress, limited the activity to the most coastal fringes and limited the productivity of trawling for white fish (Furnestin 1961). At the end of the Algerian War of Independence, the departure of maritime equipment further penalized the sector.

It was not until the 1970s that the Algerian Fisheries Office acquired 40 offshore purse seiners, simultaneously approving loans for the acquisition of ships (Baba-Ahmed 1993) so as to increase production. It also reinforced port and processing infrastructures and the training of fishermen. The sector's growth should have made it possible to rapidly double production, reaching 50,000 metric tons, particularly through the increased productivity of trawling which was previously stagnant or even possibly in decline[13]. Through the 1980s, there was strong growth in the fishing industry, the number of ships and production, which surpassed 100,000 metric tons. It was stabilized at the end of the 2000s, following plans to relaunch fishing under the auspices of its own ministry, at around 140,000 metric tons for around 4,500 ships and 40,000 seamen[14]. Catches remained dominated by oily fish, which represented around three quarters of all catches, or even more at the end of the decade. The weaker white fish production can be blamed on the poor equipment of this type of fishing, which limited its use at greater depths. Regardless, this production has tended to grow with increases in equipment in the "petit métier" sector thanks to the relaunch plans: their number having doubled since the 2000s[15]. However, fishing productivity has not progressed by much with production following in a linear manner with the increase in operations.

In El Kala, fishing has been at the root of the humid zone's habitation since the Middle Ages. Arabs have exploited coral on this site since at least the second millennium, before the Italian maritime powers developed the know-how to do so. At the end of the Middle Ages, France established a trading post in the area, which would go on to perpetuate this fishing industry. Artisanal fishing was developed particularly in the twilight of piracy in the Mediterranean. The Italian ships that lifted coral also practiced fishing so as to bring salted fish back to their Italian ports of origin. Fishing increased with the colonial conquest, throughout which the occupying power encouraged Italian boats to set up in El Kala. With a drop in demand for coral at the end of the 19th century, and thus a drop in its price, fishing production gained in importance, including inland fishing.

---

13 See: http://www.fao.org/docrep/005/d8317f/D8317F03.htm [Accessed 19 January 2023].
14 Pescamed (2013). Monde du travail, organisations des producteurs, organisations des consommateurs et formation. Report, Mediterranean Agronomic Institute, Bari.
15 *Ibid.*

16.3.1. *Lake and lagoon fishing*

Inland fishing began during colonization and was first centered around the Mellah lagoon (860ha), located between Cap Rossa and the former Calle (found at the side of the Mellah lagoon's mouth; see the EKNP map). Fishing exploitation in the lagoon began in the 1920s when a colonizer of Italian origin constructed, in the channel linking the lagoon to the sea, a system of dikes to trap fish at the mouth. This concession was equivalent to private management, as this entrepreneur was the only person authorized to exploit these fishing resources, even though they employed local hands. Since then, this exploitation has taken place through a system of dikes, catching fish as the water enters and exits, according to seasonal variations in the temperature and salinity of the water.

With independence, fishing in the El Mellah lagoon was placed under the authority of the Algerian Fisheries Office. As it opens onto the sea, this ecosystem benefits from influxes of marine biomass, in the form of numerous small fry, which relies on good management of the channel. This was thus the purview of the Office, in as much as it proved to be determinant in the maintenance of the lagoon's biological equilibrium (FAO 1982). Under the aegis of the Algerian Fisheries Office, the fishweir system was combined with net fishing within the lagoon, led with the aid of surrounding communities. With the advent of decentralization in public management, the National Algerian Fisheries Association (ENAPECHES) succeeded the Office in 1979 to better develop the lagoon and Lake Oubeira. The residents living in the douars surrounding the lake thus contributed as salaried workers in market exploitation.

At the beginning of the 1980s, production increased to at least 60 metric tons annually per body of water (FAO 1982). Development was the responsibility of the inhabitants who exploited the water with the aid of their children, while their women practiced farming and herding. The state employed at least 60 people in fish production, who, following the business' bankruptcy at the end of the 1990s, were dismissed. With their families, they turned to a commercial fishing oriented towards local markets. Through this dark decade, the nation's debt increased, given the economic difficulties that accompanied the conflict. The counterpart to the loans that accompanied the structural adjustment negotiated with international financial institutions was a vast program of privatization. Then the bell tolled for public management, but the effective privatization of inland fishing in El Kala did not arrive until a decade later. The facilities were sold as part of a monopoly on the development of the Mellah lagoon and its surrounding lakes.

Through 2005, the right to exploit three bodies of water was granted by means of state concession to a private business for a period of 25 years. It was supposed to carry out its exploitation while respecting precise regulations relating to their location in integral protected areas of the park, given their status as sites of international importance in the Ramsar convention.

With regard to the lagoon, the concessionary benefited from the know-how of the salaried workers from the earlier period. Some fishermen continued to fish with their own material (boats and nets) to resell the fruit of their expeditions to the business, which allowed it to limit its operating expenses and declare fewer catches, particularly for eel fishing. This also took advantage of export opportunities offered by the increasing rarity of the species worldwide and also locally[16]. The fishing monopoly in the park broadly pushed young douar residents to seek remunerated employment in the town of El Kala, particularly in maritime fishing.

The Oubeira and Tonga lakes are also exploited for their fishing resources. Lake Oubeira is a freshwater lake, which was also the object of exploitation under the aegis of ENAPECHES. The repopulation of fished species, to the extent that the lake does not benefit from natural influxes of marine biomass, thus fell on the business. Fishing in Lake Oubeira was carried out by salaried workers using nets, but the concession holder essentially relied on fishers living around the lake. Eel production, the lake abounding with them, was diverted in the form of exports towards Italy via Tunisia. Beside the employment of some salaried workers, the business worked with a network of fishes, who found in it a source of complementary income, despite low resale prices.

In 2009, with the implementation of the Convention on International Trade in Endangered Species of Wild Fauna and Flora (CITES), Algeria fell under the obligation to make a management plan for the exploitation of eels, considered endangered since 2007, if it wished to export them, particularly to Europe. The absence of any such management plan blocked the business' exportation opportunities, leading to its withdrawal from the Oubeira and Tonga lake concessions from 2010 (it retained the Mellah lagoon concession, in which it focuses on white fish). It has since relied on its network of fishers to allow it to purchase lake eels, some of which are sold to Tunisia with a fish health certificate, to get around the rules blocking the exportation of a species listed in the CITES.

Lake Tonga was also exploited for its fish resources, although the means of its exploitation were different from those in the other bodies of water in the park. It was never developed by ENAPECHES, and the concession accorded to the business bore only on three hectares for a period of five years. The concession thus covered only a

---

16 In Algeria, until the beginning of the 1990s, the production of the El Mellah lagoon was dominated by eels, which represented between 50% and 80% of fishing income; this proportion then dwindled, reaching a nadir of 20% at the beginning of the 2000s (Kara and Chaoui 1998).

tiny part of the northern shore of the lake. Fishing in Lake Tonga targets only eels, exploited by fishers who sold their catches to the business, at a price fixed by the latter (around €1 per kilogram), in the absence of other regular buyers. Despite the concession holder's withdrawal from the 3 ha allotted to them, the residents of the lake continued to sell the fruit of their labor to them.

Since the privatization of inland fishing in the wake of the structural adjustment, it is difficult to know the real production levels of these bodies of water. The business resorted to hiring residents, while profiting from the catches of non-salaried workers to increase its commercialized quantities beyond those declared. This system does not allow for any control over withdrawals made from the ecosystem. The impoverishment of the spaces and poor management of the Mellah lagoon channel could have affected the ecological equilibrium of these bodies of water even before the end of the concession. Despite the blocking of residents from exploiting both bodies of water, which led to fishing activity moving towards maritime areas, the sustainability of this development is seriously questioned by statements attesting to the state of the spaces by concessionary employees and residents, despite the conservation status of the bodies.

This concession structure is susceptible to lead to economic difficulties as well as environmental ones, to the extent that costs to the collective are significant, in terms of local employment as well as in terms of the future restoration of the spaces. Inland fishing has not seen the kind of development that maritime fishing has in the region, since it remained at levels equivalent or even inferior to those before its public ownership; around 60 metric tons in 2008 (Chakour 2010). But these official numbers ignore the exploitation carried out by local fishers on behalf of the business.

16.3.2. *Maritime fishing*

Maritime fishing is well established in El Kala, having been practiced since the Middle Ages, particularly the exploitation of coral. France encouraged the migration of Italian maritime populations, trying to settle them in Algeria to promote a local sector. In addition to small sailing boats introduced by the Italians, semi-industrial ships have also been deployed since the colonization. A total of 15 trawlers from the mainland were set up in El Kala (Tomas 1969). The cannery was encouraged to meet mainland demand, and also to supply non-coastal regions of the colony. These preoccupations led to the creation of a cannery at Calle and a shrimp packaging plant. This flurry of maritime activities reinforced coastal fishing, even while catches of oily fish dwindled between the wars (Tomas 1969). With independence, after the operators of the small and industrial fishing boats left, production declined significantly. Efforts to catch oily fish were relaunched, particularly through the sale of purse seine trawlers (Tomas 1969).

Today, fishing in El Kala remains coastal, even while industrial fishing survives through different operations and has, as elsewhere, profited from the liberalization of the economy and the littoralization of the last two decades. Fishing tends to be concentrated in the gulf shaped by the Roux (or Segleb) and Rossa capes. All fishing is based in the port of El Kala, the infrastructure deficit of which hinders productivity. The demands of fishers, in terms of equipment, are the consequence of the productivist approach privileged within the sector since the introduction of liberalization measures to the Algerian economy.

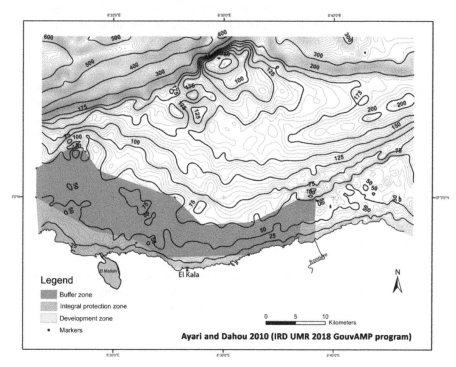

Figure 16.3. *Zoning of the future Marine Protected Area in El Kala (bathymetric chart)*

The El Tarf wilaya has seen the number of seamen and boats double in a decade, and the fishing sector has become determinant for a wilaya lacking in activities and revenue. In the EKNP, employment is rare and in large part seasonal. Even though it has not benefited from the same investment efforts as it has in other regions of

Algeia, the fishing sector has seen a strong expansion. The sector's population, averaging at a thousand seamen in the 1990s, doubled in the 2000s, following increases in fishing operations, while production has seen the same evolution – a dynamic identical at the national level (Boushaba 2008, p. 22) – passing from 1,500 to a little over 3,000 metric tons[17]. Despite this growth, the local context remains marked by a lack of investment, and the fishing boats remain dilapidated.

In the 2000s, action plans for a fishing relaunch materialized, in El Kala, in the form of financing for offshore trawlers, while financing for the purchase of fishing units at the national level privileged instead artisanal "petits métiers" that would respond to the dilapidation of the existing boats: more than 25 years old at the end of the 2000s (Boushaba 2008, p. 22). In this wilaya, both successive plans focused their encouragement on a profitable sector like trawling, rather than reinforcing the sustainability of artisanal fishing.

Despite the strong stimulation in the sector from the 1990s onwards, productivity remains low and the effort going into fishing is increasing, according to bosses in the sector, who demand longer and longer trips out to sea, and an increasing depth of catch. The fleet can be divided into a dozen trawlers, 50 sardine boats and 30 small boats kept for "petit métier fishing". Furthermore, around 300 small fishing boats are kept for coral exploitation. The entirety of the El Kala fleet is severely dilapidated. A presentation of the different forms of fishing in the region will allow for a definition of the socio-economic characteristics of the three different methods used in the maritime space of El Kala.

16.3.3. *Trawling*

Trawling activities, targeting white fish, are little developed in El Kala, given the size of the port. Furthermore, its trawlers are dilapidated, as is their equipment. El Kala's fleet is alone in trawling the gulf. Trawlers from Annaba, sufficiently motorized to fish near the coast, focus on the continental shelf (around 200 meters in depth).

This method of fishing is governed by regulations regarding the minimum size of the trawl's mesh. As elsewhere in the Mediterranean, bottom trawling, characterized by its lack of selectivity[18], is regulated because of its effects on habitats. Trawlers,

---

17 Directorate for Fishing of the El Tarf Wilaya in 2008.
18 Trawlers catch whiting, red mullets, sea bream, horse mackerel, rays, shrimp, sardines, swordfish, etc.

by dragging their nets across sea floors, affect the benthos[19] and erode the biodiversity of their fishing zones. Because of its impact on ecosystems and fish productivity, trawling is more strictly regulated in coastal areas where spawning areas can be found. Trawlers are thus not authorized to work below the alignment of capes Rossa and Roux (or Segleb); a line which corresponds approximately to an isobath of 50 m (Decree of April 24, 2004 setting limitations on the use of trawls[20]). This regulation, however, is only rarely respected, the majority of El Kala trawlers catching their haul at bottoms below 50 meters because of insufficient controls on the sea.

Those running trawlers argue based on the dilapidation of their equipment, which does not allow them to fish beyond certain distances despite the decline in productivity on closer seabeds and claim a lack of support for investment. Regulations also suggest a period of fishing further out from the coasts to allow for a kind of biological rest, given that gulfs act as reproduction zones for many species. This period of "gulf closure" – trawling is banned from May 1 to July 31 within a three nautical mile limit (Decree of April 24, 2004 setting limitations on the use of trawls) – is also ignored by actors within this fishing sector, mortgaging in the long term the renewal of fish resources.

The number of employees is usually six per trawler: a boss, a mechanic and four seamen for manpower. The profits from each journey, once exit fees have been subtracted, are shared communally, with half of all remuneration going to capital and the other half to labor. The severe exploitation of traditional seabeds and the cost of mechanized fishing limit profitability and revenue. Average individual revenue remains in the order of €400 a month, comparable to that of "petit métier" fishing, which speaks to low productivity.

Average production by unit is also shown to be low by production estimations made by fishermen, as they do not surpass 100 metric tons per year (Chakour 2010). These estimations highlight the elevated cost of boat acquisition, given the bribes necessary during allocations and speak to the absence of credit for running costs. Because of this, they have adopted a strategy of low investment and limiting trip costs by privileging sites close to the coast. The lowering of running costs allows for income to be generated more regularly and costs to be made profitable more quickly, but this is often to the detriment of the boat and to the investments needed to improve profitability. The gradual erosion of traditional fishing beds thus constitutes a limit for the growth of trawling revenue.

---

19 Species of fauna and flora affixed to the seabed.
20 Decree of 4 Rabie El Aouel 1425 corresponding to April 24, 2004.

16.3.4. Seine purse fishing

This form of artisanal fishing is also affected by dilapidation, thanks to the meager aid it has been allotted. In El Kala, seine purse fishing is carried out on old wooden boats of at least 10 meters, which are used almost exclusively to catch small pelagic[21] fish destined for local markets. Sardine boats make up the majority of the port's fleet, with around 50 units. They fish with purse seines hundreds of meters in size and in spaces usually situated in the gulf. They do not privilege particular areas of the seabed, as sardines are migratory fish. Sardine fishing is done at night, with the use of lamps which lure shoals of sardines to the surface. There are no restrictions on this type of fishing, which is selective and does not impact seabeds. This form of fishing is carried out in all zones, throughout the year.

It is nevertheless strongly marked by seasonality, sardine boats venturing out less often in winter, given the likelihood of bad weather on the sea. Currents are strong in the winter and tend to draw sardines towards the coasts, concentrating fishing efforts in the gulf, at least 50 meters' depth. Nevertheless, such activities gradually intensify beyond the gulf in the spring and in summer at depths below 50 meters.

The climate and displacement of the resource itself are determinant factors in the seasonal distribution of this artisanal activity, to which correspond catch sizes and correlated prices. The activity is split into two seasons: a low season, in which fish quantities are small but prices high, and a high season, in which quantities are large but prices very low. Production by unit and expedition varies according to these periods from a quarter of a ton to five or six tons in the most productive months, leading to large variations in price.

Sardine boats are thus highly dependent on buyers who move around and offer prices based on variations in production. The majority of catches are brought up between May and August, with prices which will decrease throughout this fishing period from 3,000 DA per 25kg box to the derisory price of 200 DA because of its abundance and the difficulties inherent in conserving this small pelagic fish through hot period, preventing transport over long distances.

This form of fishing, carried out with purse seines hundreds of meters in size, is labor-intensive, teams being made up of a dozen seamen (a captain, a mechanic, a lighting expert and seamen for manpower). The same means of splitting the profits seen on trawlers is used in purse seine fishing, as the gains derived from the activity equally remunerate capital and labor. Because of this, individual revenues are smaller, as this method requires a dozen hands, meaning a smaller division of the profits is received by each. Sardine fishers tend to agree that they are poorly

---

21 Relating to a water column unconnected to the coast or the seabed.

remunerated, invoking, on the one hand, the large remuneration received by the capital within the unit and, on the other hand, the negotiating power of the buyers.

But sardine fishers insist above all on the designation of prices to explain their low revenues. Fishmongers, few in number in El Kala, agree on purchase prices and impose them on the fishers, in the absence of a cannery to process the catches. An average revenue for a simple seaman would hover around €300 per month, below those of seamen on trawlers or "petits métiers".

16.3.5. *Drift net and longline fishing*

This form of fishing would be classed under "petit métier" fishing, given that it is practiced on small boats of seven meters, equipped with weak motors (120–140 hp), and characterized by mostly unmotorized labor.

It targets high-value species that are not caught in large quantities, given the relatively non-intensive techniques involved. We can identify around 30 small boats of this type in El Kala's port, although only a dozen are regularly used. All of the fishers of this type engage in lobster fishing, given its exportation value. From the end of April to the end of November, this type of fishing is carried out with trammel nets (nets set on the seabed) of a fairly broad mesh (at least 50 mm), positioned for three to six days on the seabed. The nets reach several hundred meters in size and are positioned 100–180 meters deep depending on the season; that is to say, beyond the gulf. They also catch white fish in these nets when the prices set by the fishmongers for lobsters are too low.

This is a form of fishing that involves a broad variety of fish, with methods which select for the targeted species according to their varying prices on the market. The fishers suffer less from the dictates of the fishmongers, to the extent that white fish are less perishable than oily fish, and their prices tend to be higher. Fishing for lobsters and white fish generally requires small teams of three to four seamen. Profits are shared in the same manner as on trawlers.

Lobster fishing is supplemented by longline fishing both on the seabed and on the surface. Deep longline fishing targets white fish, while on the surface it targets oily fish. Lines are dropped into the sea and then pulled in with fish attached by hooks. White fish hauls can reach up to 20 kilograms from a single expedition, while for oily fish, caught on drifting longlines, they are more varied, reaching up to 400 kilograms if a tuna or a swordfish is caught. However, the lower remuneration for oily fish and the difficulty involved in catching them have relegated this form of fishing, especially since the space taken up by coral fishing limits the deployment of drifting longlines. The activity is more seasonal than other forms of fishing, since poor weather is a significant constraint for such small boats. They tend to fish more from May to October.

The productivity of this form of fishing is relatively low, with an average haul of 40 kg per 12-hour excursion, each of these excursions involving both trammel net and longline techniques. Running costs, however, are not particularly high given the techniques, which leads to higher profits than purse seine fishing, once the half of the profits destined for remunerating the operation's capital are subtracted, as is the case for the other forms of fishing. It provides average monthly revenues to seamen in the order of €400, equivalent to that of the trawler seamen who target the same species.

16.4. Marine conservation and declinist rhetoric

Following this presentation of different forms of fishing, can we conclude that the production limits are attributable more to economic constraints than to fish stocks? While El Kala's fleet is not large, its modest production is above all attributable to the dilapidation of its fishing equipment, as well as to the absence of any credit apparatuses which could support investment. To this difficulty can be added the question of the designation of prices, which particularly harms sardine fishers. Profitability is thus low (Chakour 2010), even though the activity provides employment and regular income, as much for the inhabitants of El Kala as for those of the douars and small municipalities within the EKNP.

Work on fishing boats proves to be unprofitable, even though to a lesser degree in the industrial sector. Furthermore, the fact of its unprofitability is considerably more pronounced where labor hierarchies are strongly marked by differences in status, and where mobility between these statuses is hindered. These complaints are still more intense in sardine fishing, where working conditions are more severe than in other forms of fishing and revenue which are equally low. This form of artisanal fishing, the largest sector in the port by number of seamen (around 500), given the intensity of its labor, is hindered more by the disorganization of the industry (the price effect) than by any sustainability limit.

On the contrary, fishing for white fish, whether by artisanal ("petits métiers") or industrial means (trawling), seems to be approaching the problem of sustainability. While the tension between a long-term revenue for the team and the short-term profitability of the equipment for the owner is inherent across the different forms of fishing, white fish fishing is confronted more specifically with the problem of its resource's apparent decline. In El Kala, pelagic fishing represented two thirds of all landings in the 1990s, reaching more than three quarters in the 2000s, while at the same time the number of sardine boards remained stable and the number of "petit métier" boats tripled (the number of trawlers also remained stable). This evolution can be linked to the gradual decrease in productivity of traditionally targeted seabeds in the gulf, as we can read in the documents aimed at the conservation of the EKNP's maritime area (Bouazouni 2004; Grimes 2005).

Public policies limit exploitation through legal apparatuses that are supposed to guarantee the renewal of the resource in question and at the same time encourage exploitation through direct (financing the acquisition of boats) or indirect (fuel costs are an important factor for a form of fishing as thoroughly motorized as trawling) subsidies, or even through a deficit with regard to the control of activities on the sea. While responsible fishing standards are inscribed in the Fishing Code, they are seldom applied. In their analysis of the low productivity, fishers first blamed the lack of respect for fishing zones and the biological renewal period by industrial fishing outfits, which tend to constantly target the same seabeds in the El Kala gulf. While an increase in fishing units is evident, the fleet remains artisanal, and its dilapidation limits activity to only the most coastal areas, already intensely exploited, particularly for white fish species. Fishers note a continued weakening of production per unit. They are facing increasing running costs with increases in the effort necessary to catch the fish and insist on the shortcomings of public authorities with regard to the sector's regulation, despite the heavy taxes already imposed by the state.

Despite the low intensity of El Kala's fishing efforts on the whole, its concentration within the gulf and the increase in units foreshadow damage to the ecosystem. This poor distribution of fishing efforts seems to be the source of the dwindling number of fish. A better distribution of fishing efforts could be prompted by more precise planning than the present in the Fishing Code, as well as by a reinforced application of the latter, given the illegal practices at play. These measures are being overlooked, however, in favor of the project of extending the park to encompass its maritime area. The maritime part of the EKNP is destined to become a Marine Protected Area. However, the extension of the coastal park across its maritime area is only partially justified by the intensity of fishing.

Despite the public discourse around the site's conservation, as promoted by the Ministry for Agriculture, in charge of national parks, and the Ministry for the Environment, in charge of coastal conservation, the Ministry for Fisheries throws some cold water on the diagnosis of a decline in fish resources that is being taken to justify the site's conservation. Behind the simple fact of the sector's poor economic performance, different discourses that are part of the controversies surrounding public action are able to hide. A productivist perspective is borne by the Fisheries Administration, whose successive relaunch plans endorse a vision of potential growth not yet attained, and attainable through the pursuit of public investment. Opposed to this vision is a discourse quietly supported by a conservationist understanding of fish resources, which insists on the goal of limiting catches within the gulf, given the concentration of fish in these coastal areas, and making the gulf a Marine Protected Area to allow for the restoration of fish stocks.

However, we can understand the hypothesis of the decrease in white fish fishing productivity differently, as white fish have fallen from comprising a third of all catches in the 1990s to only 10% in 2008. Coral fishing is never mentioned in statements about the decline in catches. While coral fishing has been banned since 2001, its sale as contraband has seen an exponential growth throughout the last decade, which could be the source of a move among "petit métier" fishers towards coral fishing. Despite testimony to this end from fishers, no official expert on the area has pointed out this fact. The increase in fishing boat registrations, above all with regard to "petits métiers," is without doubt primarily with an eye to coral fishing, given the stagnation in general of white fish production.

These elements are more likely to disrupt the productivist approach of fishing services than the ecosystem restoration strategy, stemming from the hypothesis of resource decline formulated by conservation administrations, to the extent that this black market does considerably more harm to ecosystems than fishing. Public discourses diagnose the stagnation in catch numbers in the same way, but are opposed to each other with regard to remedies according to their individual interests. The action of the state is split between two administrations, with the Fishing Directorate, on the one hand, promoting the productivist approach, and Park Management, on the other hand, favoring the conservationist angle.

16.5. Conclusion

In this context, can we not identify a declinist discourse running through conservationist perspectives on fish resources, solely to impose certain public interests at the expense of other sectors within the administration? At the same time, does this not mask the exclusion of certain parts of the artisanal fishing sector to the profit of coral fishing? In consequence, ought we to conclude with a genealogy of these declinist discourses, from those originating during the colonial period of forestry management (which reinforced the prerogatives of forestry officials and the interests of concessionaries) to those present in contemporary maritime fishing?

Since the colonial period, marine conservation in Algeria has been bound up in an approach favoring an increase in fishing productivity. Contrary to certain other works which have not detected the dimension of market development present in colonial conservation policies (de Peyerimhoff et al. 1937), their close ties are evident in the case of Algeria. Economic lobbies were active in the formulation of forest and water policies. With regard to fishing, it was a matter of making room for ships from the mainland and ensuring the renewal of the fish stocks targeted, with the goal of intensifying their exploitation. Varied protection measures were applied across the territory, limiting mesh sizes, fishing periods and fishing zones. Today, while this vision is still present, declinist arguments are being put forward, tied to general conservation programs. They justify the prohibition advocated by the Park's management, which itself displays in a more evident way the legacy of declinist discourses stemming from colonial forestry. This grafting of contemporary biodiversity conservation paradigms onto declinist theories is significant on land and asks the question of a possible convergence on maritime spaces.

We can see the ambiguities behind the controversies, with regard to ecosystem knowledge and local uses of natural resources, as much on land as on the sea. Behind the affirmation of a conservationist vision, ways of excluding certain uses – pastoralism on land and artisanal fishing in the sea – to the profit of models that grant concessions to private businesses are hidden. This public discourse, based on the very selective use of scientific knowledge, or even its omission, never questions concessionary models, whether with regard to forestry businesses or inland fishing businesses. We can see the same ambiguity at play in the maritime part of the park, where the management plan for the future Marine Protected Area is likely to considerably affect the artisanal sector (Dahou 2018), while ignoring the problem of illegal coral fishing.

16.6. References

Agrawal, A. (2005). *Environmentality. Technologies of Government and the Making of Subjects*. Duke University Press, Durham.

Aubréville, A., Babbey, A., Barclay, E.-N. (1937). *Contribution à l'étude des réserves naturelles et des parcs nationaux*. Lechevalier, Paris.

Ben Hounet, Y. (2013). Propriété, appropriation foncière et pratiques du droit en milieu steppique (Algérie). *Études rurales*, 192, 61–77.

Berlan-Darqué, M. (ed.). *Histoire des parcs nationaux. Comment prendre soin de la nature ?* Quæ, Versailles.

Blanc, G. (2015). *Une histoire environnementale de la nation. Regards croisés sur les parcs nationaux du Canada, d'Éthiopie et de France*. Publications de la Sorbonne, Paris.

Bouazouni, O. (2004). Parc National d'El KALA. Étude socio-économique du PNEK. Report, Projet régional pour le développement d'aires marines et côtières protégées dans la région de la Méditerranée (MedMPA). PAM, PNEK, PNUE, RAC-SPA.

Boushaba, A. (2008). L'Algérie et le droit des pêches maritimes. PhD Thesis, Université Mentouri, Constantine.

Chakour, S.C. (2010). Présentation du secteur de la pêche à El Tarf. GEMALIT, Programme GouvAMP WP1 – 2011. Report, Rentabilité de la pêche, rente halieutique et marché du poisson à El Kala.

Cutler, B. (2014). Water mania! Drought and the rhetoric of rule in nineteenth-century Algeria. *The Journal of North African Studies*, 19(3), 317–337.

Dahou, T. (2018). *Gouverner la mer en Algérie. Politique en eaux troubles.* Karthala, Paris.

Dahou, T. and Wedoud Ould Cheikh, A. (2007). L'Autochtonie dans les Aires Marines Protégées. Terrain de conflits en Mauritanie et au Sénégal. *Politique africaine*, 108, 173–190.

Dahou, T., Weigel, J.Y., Saleck, A.M.O., Simao Da Silva, A., Mbaye, M., and Noël, JF. (2004). La gouvernance des aires marines protégées : leçons ouest-africaines. *VertigO*, 5(3).

Davis, D.K. (2007). *Resurrecting the Granary of Rome. Environmental History and French Colonial Expansion in North Africa.* Ohio University Press, Athens.

De Belair, G. (1990). Structure, fonctionnement et perspectives de gestion de quatre écocomplexes lacustres et marécageux (El Kala, Est algérien). PhD Thesis, Université des Sciences et Techniques du Languedoc, Montpellier.

FAO (1982). Analyse et développement d'une exploitation lagunaire intensive. Exemple du lac Mellah – Algérie. Projet régional de développement de l'aquaculture en Méditerranée. 1995, Code de conduite pour une pêche responsable. Report, FAO.

Ford, C. (2008). Landscape conservation, and the anxieties of empire in French Colonial Algeria. *The American Historical Review*, 113(2), 341–362.

Furnestin, J. (1961). La pêche maritime en Algérie et ses possibilités. Rapport de mission. *Revue des travaux de l'institut des pêches maritimes*, 25(1), 21–32.

Grimes, S. (2005). Plan de gestion de l'aire marine du parc national d'El Kala. Report, MedMPA, PNUE, PAM, CAR-ASP.

Guignard, D. (2010). Conservatoire ou révolutionnaire ? Le sénatus-consulte de 1863 appliqué au régime foncier d'Algérie. *Revue d'histoire du XIXe siècle*, 41, 81–95.

Hachemaoui, M. (2013). *Clientélisme et corruption dans l'Algérie contemporaine.* Karthala, Paris.

Harper, M. and White, R. (2012). How national were the first national parks? Comparative perspectives from British settler societies. In *Civilizing Nature. National Parks in Global Historical Perspective*, Gissibl, B., Höhler, S., Kupper, P. (eds). Berghahn, New York.

Homewood, K. (1993). Livestock economy and ecology in El Kala, Algeria: Evaluating ecological and economic costs and benefits in pastoralist systems. Pastoral development network paper n.o. 35a. Report, Overseas Development Institute, London.

Jentoft, S., Christiaan van Son, T., Bjørkan, M. (2007). Marine protected areas: A governance system analysis. *Human Ecology*, 35, 611–622.

Jones, K. (2012). Unpacking Yellowstone: The American national park in global perspective. In *National Parks in Global Historical Perspective*, Gissibl, B., Höhler, S., Kupper, P. (eds). Berghahn, New York.

Kara, H. and Chaoui, L. (1998). Niveau de production et rendement d'une lagune méditerranéenne : le lac Mellah (Algérie). Report, Commission Internationale pour l'Exploration Scientifique de la mer Méditerranée, 35.

Larrère, R., Lizet, B., Berlan-Darqué, M. (eds) (2009). *Histoire des parcs nationaux. Comment prendre soin de la nature ?* Quæ, Versailles.

Mathevet, R., Thompson, J., Delanoë, O., Cheylan, M., Gil-Fourrier, C., Bonnin, M., Mathevet, R. (2010). La solidarité écologique : un nouveau concept pour une gestion intégrée des parcs nationaux et des territoires. *Natures sciences sociétés*, 4(18), 424–433.

de Peyerimhoff, P., Fairhead, J., Leach, M., Scoones, I. (eds) (1937). Green grabbing: A new approriation of nature? *Journal of Peasant Studies*, 39(2), 237–261.

PNEK (2011). Actualisation du zonage du parc national d'El Kala. Phase 1. Parc national d'El Kala. Report, PNEK.

Selmi, A. (2009). L'émergence de l'idée de parc national en France. De la protection des paysages à l'expérimentation coloniale. In *Histoire des parcs nationaux*, Larrère, R., Lizet, B., Berlan-Darqué, M. (eds). Quae, Versailles.

Simonnet, R. (1961). Essai sur l'économie des pêches maritimes en Algérie. *Revue des travaux de l'institut des pêches maritimes*, 25(1), 33–124.

Tomas, F. (1969). Annaba et sa région agricole. *Revue de géographie de Lyon*, 44(1), 37–74.

Walley, C.J. (2004). *Rough Waters: Nature and Development in an East African Marine Park*. Princeton University Press, Princeton.

West, P., Igoe, J., Brockington, D. (2006). Parks and peoples: The social impact of protected areas. *Annual Review of Anthropology*, 35, 251–277.

Zytnicki, C. (2013). Faire l'Algérie agréable. Tourisme et colonisation en Algérie des années 1870 à 1962. *Le mouvement social*, 242, 97–114.

17

Social Representations of Biospheres and Sustainable Local Development in Bou Hedma (Tunisia)

17.1. Introduction

This study focuses on social representations of protected areas in Tunisia and their role in sustainable local development, more particularly with regard to the living conditions of those living near these spaces. It aims to better understand the social relations enjoyed by local communities with their area. The objective of this work is to show the importance of studying social representations as a key part of understanding the expectations of the relevant actors and consequently of understanding the social acceptance of a protected area. This study is based on Bou Hedma Park in central Tunisia, which is representative of other national parks.

Through this research, we seek to identify the social representations of the park formed by the inhabitants of Bou Hedma, so as to clarify their understanding of its importance to sustainable local development, particularly with regard to the protection of biodiversity and to the relationship between local people and their environment. This exploratory study was carried out following an essentially qualitative approach, based on participant observation, a questionnaire and semi-structured interviews.

The study and the analysis of social representations are merely explications of the gaze leveled at a particular object – here Bou Hedma Park – with which local people associate a wide range of realities. These realities are determined by the age, the level of education and the attitude to the reserve of these local people. It is

---

Chapter written by Abdelkarim BRAHMI.

important to take these social representations into account if we wish to understand and explain the practices and attitudes of those living in the region towards the park, to find solutions for the protection of the park and to plan programs and outline strategies involving different actors, particularly local people and civil society. Consequently, the study of the locals' social representations of Bou Hedma park is highly important for the implementation of programs, which could have positive effects on the region's population. Through the intermediary of social representations, it becomes possible to understand the meanings and values accorded to the park.

After a brief presentation of our area of study, we will introduce the data analysis and methodological aspects of the investigation. Next, we will present the results of our study and outline a discussion around how we can interpret the different social representations encountered therein.

17.2. Bou Hedma National Park

Bou Hedma National Park, created in 1980 by Presidential Decree no. 1660, was listed as a UNESCO Biosphere Reserve in 1977, within the framework of the Man and the Biosphere (MAB) program. Located between parallels 34°27' and 34°32'N and meridians 09°23' and 09°41'E, it covers an area of 16,500 hectares (Karem et al. 1993), shared between two governorates (Sidi Bouzid and Gafsa) and is found in central Tunisia, 107 km to the north-west of the town of Gabès, 60–75 km to the east of the town of Gafsa, 45 km to the north-west of the town of Skhira and 120 km south-west of the town of Sfax (Tarhouni et al. 2007).

It is classified as within the arid bioclimatic stage (Chaieb and Boukhris 1998), with an average annual rainfall of around 180 mm (Noumi 2010) and dry periods which can last up to 10 months per year (Karem et al. 1993). The average temperature in the coldest month hovers between 3°C and 4°C and that of the hottest month between 30°C and 35°C (Le Houérou 1995). Temperatures can sometimes reach down to –4°C (Zaafouri et al. 1997). Classified as a steppe or semi-desert space, covered in extensive herding grounds, the park offers different kinds of spaces including mountainous massifs, intermediate shelves, plains and thalwegs (Dalhoumi et al. 2015).

The biosphere reserve is divided into three areas: Bordj Bouhedma, Haddej and Belkhir (Jaouadi et al. 2012). It plays a decisive role in preserving biodiversity, thanks to its biological wealth. It offers refuge to many animal species, some of which having been reintroduced and some being on the road to extinction. It has a broad range of flora, characterized by the domination of the *Acacia tortilis raddiana* (gum tree) forest, the only one in Tunisia. For this reason, Bou Hedma constitutes a

North African pseudo-savannah landscape (Zaafouri et al. 1997), representative of the Sahel region and almost nonexistent elsewhere in North Africa.

Figure 17.1. *Bou Hedma National Park (source: Google). For a color version of this figure, see www.iste.co.uk/romagny/biosphere2.zip*

17.3. Methodological research framework

Our field study was carried out in August 2020. The interviewed population was part of those living around the park. The study relied on a sample of 200 people, split into two equal groups based on age. The first group encompassed those over 40 years old. The second group consisted of educated young people between 20 and 40 years old. Twenty people were interviewed, drawing evenly on the two age groups.

The collection of data relating to social representations of the Bou Hedma reserve is based on participant observation in the park, as well as in neighboring locales, on the semi-structured interviews and on the questionnaire. These three data collection tools seem to us to be complementary. Participant observation is "indispensable to a qualitative research approach" (Ouellet 1994), for completion of the information obtained through the interviews and the questionnaire, as well as through written sources. It allows us to identify both the daily reality of those living in Bou Hedma and the reality of the park and its role in sustainable local development.

Alongside participant observation, questionnaires often represent the first level of data collection (Maury 2007). It is currently the most important and most widely used technique in the study of social representations. It allows, on the one hand, for "the introduction of fundamental quantitative elements into the social aspect of a representation" (Abric 2011), and, on the other hand, for the characterization of the represented object (Abric 2011) through the standardization of the gathered data, the questions being the same for everyone participating in the questionnaire (Bonardi and Roussiau 1999; Mulkay 2006). Within the context of this research, the discourse corpus is made up of the inhabitants of Bou Hedma's responses to questions concerning the reserve and its role in sustainable local development.

In the second phase of the study, semi-structured interviews were conducted with 20 inhabitants (10 per subgroup). These interviews took place across the three sections of the park (Jaouadi et al. 2012), Bou Hedma el Borj, Haddéj and Belkhir, and in neighboring locales. Their theme was the Bou Hedma National Park and its role in sustainable local development. This tool required interviewers to be trained, so that the information would come only from the interviewee, without the active participation of the interviewer (Bonardi and Roussiau 1999). Semi-structured interviews aimed to promote the production by the respondent of a discourse on a theme defined within the framework of a research project. This allows us to better understand the opinions of the inhabitants of Bou Hedma with regard to the park, and the way in which they judge their role in sustainable local development.

Thus, like many other researchers, we supposed that this research tool was necessary to fill out the information gathered by the questionnaire (Blanchet and Gotman 2010; Grawitz 2010), as it allowed us to deepen the results offered by the questionnaire. Furthermore, "the cross-referencing of data from interviews and questionnaires is desirable and highly recommended" (Grenon et al. 2013). Indeed, it is highly useful to associate questionnaires and semi-structured interviews in studies of social representations. This qualitative methodology supports an in-depth analysis of the interviewees' discourses, to the extent that it allows for a gathering of participants' perceptions, beliefs and attitudes (Lemay et al. 1999).

17.4. Social representations of Bou Hedma National Park among the surrounding population

A lexical reading of the architecture of the data collected through the questionnaire filled out by those living around the Bou Hedma National Park allows us to distinguish two categories of social representations regarding the park and its role in sustainable local development, the most significant in the framework of our study. We will first consider negative representations and then positive representations.

17.4.1. Negative representations: "the park as problem"

This category of negative social representation of the park revolves around the term "problem". It occupies a central space in the representations of the park constructed by the older contingent of Bou Hedma's inhabitants. In this category, other words emerge, essentially structured around the term "problem", and which are similar and associated with one another. These include "curse", "catastrophe", "indignation" and "affliction". The questionnaire shows that this category of representation implicates 80% of the respondents from the older group, while positive social representation, which makes the park a symbol of identity, an icon, a personality and a part of their heritage, implicates only 20% of that older group. These social representations can be explained by the negative impact of the park on the daily life of Bou Hedma's inhabitants since its creation by the state in 1980.

These different social representations are based on the park's utility, particularly if we take into account the pastoral way of life of those living around it, and the vast area it covers. These lands were previously used by the inhabitants, even though they belonged to the state. The creation of the park thus prevented the use of enormous ranges of herding land.

Our analysis of the gathered date shows that 90% of the older group of participants in the questionnaire limited the park's role in sustainable local development to its economic and social dimensions. This understanding thus occupies a central space in social representations of the park's role. It also encompasses terms tied to the economic and social dimension of sustainable development, such as "employment", "fixed income", "economic investment", etc. These representations are based on the presumption that everything has a utility, and that we can derive economic, material and practical benefits from the park. Consequently, the majority of the older inhabitants of the Bou Hedma region tend to emphasize the possibility of a good economic use of the park, through ecotourism or alternative tourism. Indeed, the other aspects of sustainable development hold no interest. They are neglected by the majority of the older group of respondents, particularly the environmental aspect that spurred the park's creation in the first place. For the majority of inhabitants, the park's role in economic and social development in the region is highly limited or even nonexistent.

Alongside these negative representations of the Bou Hedma National Park, our examination of the responses of those living around the park led us to hypothesize the existence of another, very different, category of representations: positive social representations.

17.4.2. Positive representations: "the park as symbol of identity"

These "positive" social representations make the park a "symbol of identity". They make up another category of social representation shared by 75% of the younger group of respondents from Bou Hedma. For these young people, the park is also "an icon", "natural heritage" and "a personality". While negative social representations from the "park problem" category implicate less than 25% of the younger group of respondents, this type of social representation reveals the pride of the Bou Hedma region's young people in the Mediterranean and international influence enjoyed by the park. While the majority of inhabitants do not profit economically from the park, due to its marginal economic use, this second category is based instead on symbolic values.

The questionnaire shows that 90% of the younger group of respondents emphasizes the environmental role of the park in sustainable local development. This representation is therefore structured around the environmental dimension of the park. It covers several terms, including "environmental conservation", "preservation of natural heritage", "protection of the ecosystem", "preservation of biodiversity" and "protection of endangered plant and animal species". This environmental dimension is essential to this category of representation. The other representations – such as "employment", "fixed income", "economic investment", etc. – are distant to the majority of the young population, thanks to their marginal role in sustainable local development.

17.5. Discussion and interpretation

17.5.1. Negative representations: "the park as problem"

These representations, which mark the collective opinion of the majority of the older group, are the result of two factors: the negative repercussions of the park on the life of those living around it and its marginal role in the economic and social development of the region.

17.5.1.1. The negative repercussions of the park on the lives of those living around it

The foundation of these negative repercussions can be traced back to the creation of the park in 1980, a decision imposed by the state without any consultation of the local population. This exclusion explains the current attitude of the older inhabitants of the region towards the park, being the generation who lived through its creation. This unilateral decision on the part of the state during the era of Habib Bourguiba is explained by the authoritarian political system dominant in Tunisia that lasted until the revolution of December 17, 2010 and the absence of civil society. The state did not then taken into account the socio-anthropological realities of this region, an arid area characterized by a pastoral way of life.

Before the park's creation, the inhabitants of Bou Hedma exploited the vast ranges of herding land, which belonged to the state. This exploitation took different forms, including extensive agriculture (the primary economic activity and revenue source), extensive breeding (particularly of sheep and goats), the uprooting of trees, particularly *acacia tortilis*, for their wood, and game hunting, undertaken particularly in the dry season to provide for food requirements. The local population used the natural resources available in the area for their subsistence and according to tradition.

The creation of the park was accompanied by the closing-off of these natural resources, depriving the local populations of their ability to use the land. The reduction of herding ranges and the prohibition of grazing left herders unable to meet the food needs of their livestock. They made recourse to supplements of concentrate feed and barley (Roselt/OSS CT7 2004, p. 98), which obliged herders to bear the extra costs. This prohibition of the use of pastoral resources led to an acceleration of agricultural intensification in the region. Prohibitions on hunting, which aimed to protect wild fauna, were felt by those in the region as a form of "food discrimination, especially in the dry period" (Roselt/OSS CT7 2004, p. 99). Prohibitions on gathering wood deprived locals of a major source – the *Tortilis raddiana* acacia – which constituted their primary energy source. Research conducted in the countryside of southern Tunisia estimated wood consumption per inhabitant to sit at around 1.5 kg per day (Roselt/OSS CT7 2004).

Currently, wood consumption is dwindling in relation to an increase in butane consumption. But it is still high compared to other regions of the country, particularly coastal area. The wood is transformed into charcoal and used particularly for the preparation of tea, the cooking of meals and the heating of homes in winter. This explains the lack of respect by locals of the wood-gathering prohibition and reports of a great number of logging offences regarding the *Tortilis raddiana* acacia. To compensate for the deprivation of access to natural resources, the park employs a large number of workers, recruited from the regions' inhabitants, on state worksites called "hadhira" (Roselt/OSS CT7 2004). However, the majority of inhabitants reject this work and consider themselves as victims of the park's creation. This discontent explains why the park was attacked during the revolution of December 17, 2010, during the period of state weakness. Threatened animals were hunted and available resources were overused.

17.5.1.2. *The marginal role of the park in the region's economic and social development*

Bou Hedma National Park benefits from significant touristic potential tied to its natural and cultural heritage. It is distinguished by the presence of the Sub-Saharan savannah, dominated by the gum tree (*Acacia tortilis raddiana*), which makes the Bou Hedma jebel one of the last savannahs in North Africa, a symbol of the park. Development and good marketing could attract local and foreign tourists, particularly nature lovers.

Alongside this natural potential, the region has rich cultural potential: prehistoric sites, Roman ruins, marabouts, art and the popular traditions of the inhabitants, including customs, culinary arts, heritage costumes, traditional trades, popular poetry, horse-riding spectacles, etc. However, the touristic exploitation of this cultural heritage would require development of the park and the surrounding locales, in order to integrate them into the touristic circuit of park visits, and allow for a true integration of local populations. Thus, "tourism in national parks is called on, in service of economic and social development and of the conservation of natural heritage" (Souissi 2008).

However, this significant touristic potential has not been developed. This explains the marginal role of the park in the tourism sector, a finding that applies equally to other parks in Tunisia, which do not attract significant quantities of local or foreign tourism. This marginality with regard to economic and social development has been present since these parks' creations. This is due to several factors, the most important being a lack of interest on the part of the state and especially on the part of those directly involved in the ecotourism sector (Souissi 2008), as well as the exclusion of the local population from contributing to sustainable local development and profiting from it.

We cannot speak of sustainable development before the revolution of December 17, 2010, particularly in its political dimensions (transparency, participation, coordination, responsibility and governance), due to the tyrannical authoritarianism of the political regime and the absence of civil society. Sustainable development demands public participation in the different stages of different projects, from decision-making to the implementation and tracking of those decisions. This ensures transparency, integrity and governance, contributing to these projects' success. Any project that does not serve the interests of the local population is dropped and stands as a failure.

This is applicable to Bou Hedma. The inhabitants of the region did not participate in the decisions surrounding the creation of the park. Neither were they involved in its management nor in its supervision, hence the poor management of the park, and the indifference of its older inhabitants to the preservation of its animals and its vegetation, particularly the gum tree (*Acacia tortilis raddiana*), which symbolizes the park. The park was not created with sustainable local development or any improvement to the populations living conditions in mind. Rather, it was an expression of the Tunisian state's engagement with international institutions to obtain financial aid. This situation is implicated in the same development scheme that has dominated the country from the colonial period to the present day.

The marginal role of the park in the Bou Hedma region's economic and social development has continued even after the Tunisian revolution, due to conflicts of use between the different actors within the park, as shown by the semi-structured interviews. These conflicts of use involve local and regional actors, on the one hand, and the managers of the park (under the aegis of the General Directorate for Forests) on the other hand. The local and regional actors, whether public or private, seek most often to develop natural and cultural tourist activities based on the effective participation of local populations, so as to allow them to benefit from its economic consequences, which would be equitably distributed (Souissi 2008). The managers of the park consider its primary function to be the preservation and conservation of its natural heritage and biodiversity, as well as the development of their environmental aspects. To this end, they focus on visits from students and researchers and participate in educational and scientific activities aimed at protecting the park. They do not wish to establish any intensive tourist activity within Bou Hedma National Park, or in any other Tunisian national parks, as they wish to avoid the negative impact they could have on the natural resources available therein (Souissi 2008). Tourist activities in the park are indeed marginal, but they remain natural in focus, based on the contemplation of animal and plant species in their natural environments, as well as on profiting from the charm and beauty of the park's nature.

The majority of the park's older population limits its role in local sustainable development to its economic and social aspect, particularly the guarantee of stable work and fixed income.

The devaluing of the park's role and its restriction to its economic dimension are due first to the complexity of the concept, as well as to the high rates of illiteracy among the older population (Roselt/OSS CT7 2004, p. 94). Furthermore, before the Tunisian revolution, its political dimension was almost entirely absent, as the state imposed the park's creation on the population without consulting them. Moreover, the two other dimensions of the park – cultural and environmental – are neglected, and are useless to the older population, bringing them no direct material benefit.

17.5.2. *Positive representations: "the park as symbol of identity"*

Positive representations of the park can be found among the young population of Bou Hedma, who are mostly educated. They rest on the symbolic values arising from the park's importance to its young inhabitants. It is indeed thanks to the park that the region is known around the world, particularly in the Mediterranean world and above all after its designation as a UNESCO Biosphere Reserve in 1977. These representations began to appear after the Tunisian revolution on December 17, 2010, which led to a transition towards democracy. It allowed the region's young people to contribute to the management and preservation of the park through the important role of civil society. Contrary to this, the older population of Bou Hedma consider these symbolic values to signify little, in that they contribute neither to the improvement of their living conditions nor to the social and economic development of the region.

17.6. The cultural dimension

The cultural role played by the park in sustainable local development, associated with the positive representations of the park's young inhabitants, is centered around its symbolic values and is manifested in a feeling of pride at belonging to the Bou Hedma region as well as to Tunisia, despite their marginalization and poverty. This feeling is implicated in the context of an awakening of locals to their identities in a form of resistance against the influence of cultural globalization, a true threat to the different aspects of identity that directly and profoundly affect cultural specificity and personality.

This cultural dimension of sustainable local development highlights the close links between culture and nature in the daily life of those living in the park (Soini and Dessein 2016). These two types of heritage are thus complementary and integrated with each other, to the extent that it is difficult to separate the one from the other.

17.7. The political dimension

According to these positive representations, the political role of Bou Hedma National Park in sustainable local development appears to manifest in the participation of the young population and of civil society in the preservation and management of the park, and in all of the events organized within the park and its surrounding locales. This participation seeks to ensure the transparency and integrity of the park's management, and the establishment of a mutual trust between the state and the different actors involved in its management and preservation. Such notions lead to good governance. Young people's acceptance of participation in the management and participation of the park can be seen as an expression of their increased interest, which constitutes, on the one hand, a mechanism for resisting the politics of marginalization and exclusion and, on the other hand, a tool for breaking the region out of its isolation.

These different notions, belonging to the political lexicon, have a positive charge, which has given Bou Hedma's young population a feeling of citizenship, a feeling which is absent in the majority of the region's inhabitants, as in the other regions of central Tunisia before the revolution.

17.8. The environmental dimension

These positive social representations, making up the heart of the younger group's representations, are focused on the environmental role of the park, the principal factor in its creation. For the majority of young people, its role is limited to protecting the environment and preserving the region's natural heritage. More precisely, however, the park's essential environmental role comprises the preservation of biodiversity in an arid steppe, a biodiversity owed to the area's biological wealth. The park's flora is highly diversified, counting more than 500 plant species, eight of which are recognized as of principal importance in the protection of biodiversity by the National Study for Biological Diversity: *Acacia tortilis raddiana, Juniperus phoenicea, Pistacia atlantica, Thymelia sempervirens, Tetrapogon villosus, Tricholaena teneriffae, Cenchrus ciliaris* and *Digitaria nodosa* (Dalhoumi et al. 2015).

The *Acacia raddiana* (the gum tree) is the most important plant species in the park. Consequently, the park's gum forest, which constitutes a North African pseudo-savannah landscape (Zaafouri et al. 1997), is unique. The area's fauna comes from the Saharan, semi-desert and arid regions. It is highly varied and includes native and non-native species (Roselt/OSS CT7 2004). The park's importance resides in its possessing rare and endangered animals, and others that have disappeared and been reintroduced, such as the addax, the oryx and the Dama gazelle.

This park thus represents the last relic of a landscape that has otherwise disappeared from North Africa, with Sub-Saharan plant and animal species. For this, it bears an exceptional importance in the preservation of biodiversity, which constitutes one of the principal factors in the maintenance of the ecological equilibrium.

This study of the park's role in sustainable local development according to the majority of the younger population of the Bou Hedma region shows the domination of the park's environmental dimension, which, according to the majority of these younger people, is its only concrete and visible dimension, alongside its marginal economic and social dimensions. With regard to its political and cultural dimensions, despite their presence in the questionnaire and the semi-directed interviews in the form of several terms such as "participation", "implication", "civil society", "transparency", "integrity", "responsibility", "mutual trust", "governance" (the political dimension), and "feelings of pride", "symbol of identity", "icon" and "personality" (the cultural dimension), they were not well defined nor even mentioned. Thus, we can suggest that the majority of the younger population of Bou Hedma was not aware of these dimensions, despite their education, which reveals the complexity of the concept of sustainable local development.

17.9. Conclusion

This study of social representations has used a range of research tools, including participant observations, questionnaires and semi-directed interviews. This methodology has allowed for a deep study of the discourses and social representations of different actors. It has also allowed for the comprehension of the attitudes, perceptions, beliefs and values of the interviewed population, which form their understanding of protected areas. Thanks to their importance for sustainable local development, and in particular their environmental role, this study has focused on the social representations of the Bou Hedma National Park created by those living around it. It has highlighted the age distinction made between younger and older inhabitants. Finally, we have sought to identify the place occupied by the notion of sustainable local development in these representations of the park with regard to its preservation and to the improvement of the living conditions of those living around it. The identification of the nature of these representations is important to the extent that it allows us to recognize the opinions and attitudes of the inhabitants in order to ensure the acceptability of the project, which has contributed to its success.

We can observe an important difference between the social representations of Bou Hedma National Park created by the older inhabitants and the younger ones: the older inhabitants are mostly tied to negative typological categories, considering the park as "a problem". The younger inhabitants are mostly tied to positive categories, considering the park "a symbol of identity".

The concept of sustainable local development remains ambiguous for the majority of the participating population: while the older population reduces the park's role to its economic and social aspects, despite them being marginal, the young population insists on its environmental aspect, alongside its economic and social aspects. Two other dimensions of local sustainable development, the political and cultural dimensions, are completely absent from the representations created by the older population and are blurred in the representations created by the younger population.

17.10. References

Abric, J.C. (2011). *Pratiques sociales et représentations*. PUF, Paris.

Bernard, S. (2009). *Du tourisme durable au tourisme équitable. Quelle éthique pour le tourisme de demain ?* De Boeck Supérieur, Brussels.

Blanchet, A. and Gotman, A. (2010). *L'enquête et ses méthodes. Entretien*. Armand Colin, Paris.

Bonardi, C. and Roussiau, N. (1999). *Les représentations sociales*. Dunod, Paris.

Chaieb, M. and Boukhris, M. (1998). *Flore succincte et illustrée des zones arides et sahariennes de Tunisie*. L'Or du Temps, Tunis.

Dalhoumi, R., Aissa, P., Aulagnier, S. (2015). Cycle annuel d'activité des chiroptères du parc national de Bou-Hedma (Tunisie). *Revue d'écologie*, 70(3), 261–270.

Dany, L. (2016). Analyse qualitative du contenu des représentations sociales. *Les représentations sociales*.

Desrochers, V., Ferraris, J., Garnier, C. (2014). Étude des représentations sociales d'un site classé aménagé : application au site de l'Anse de Paulilles (France). *VertigO*, 14(1) [Online]. Available at: http://vertigo.revues.org/14747 [Accessed 13 October 2020].

Grawitz, M. (2010). *Méthodes des sciences sociales*. Dalloz, Montrouge.

Grenon, V., Larose, F., Carignan, I. (2013). Réflexions méthodologiques sur l'étude des représentations sociales : rétrospectives de recherches antérieures. *Phronesis*, 2(2), 43–49.

Jaouadi, W., Mechergui, K., Gader, G., Larbi Khouja, M.A. (2012). Dynamique de l'occupation des sols dans le parc national de Bouhedma en Tunisie. *Forêt méditerranéenne*, 33, 4.

Karem, A., Santini, M.K., Schoenenberger, A., Waibel, T. (1993). *Contribution à la régénération de la végétation dans les parcs nationaux en Tunisie aride*. Imprimerie Arabe de Tunisie, Tunis.

Le Houérou, H.N. (1995). Bioclimatologie et biogéographie des steppes arides du nord de l'Afrique. Report, CIHEAM, Options méditerranéennes, Montpellier.

Lemay, A.M., Nieuwenhuyse, H.V., Cottinet, S. (1999). Les représentations sociales de l'avenir chez les jeunes Québécois. Report, ERE Éducation, Université Laval, Quebec.

Maury, C. (2007). Les représentations sociales : boîte à outil. *Revue de littérature* [Online]. Available at: http://www.knowandpol.eu/.

Mulkay, F. (2006). Les représentations sociales : étudier le social dans l'individu. *Les cahiers internationaux de psychologie sociale*, 61, 57–62.

Noumi, Z. (2010). *Acacia tortilis* subsp. *raddiana* en Tunisie pré-saharienne : structure du peuplement, réponses et effets biologiques et environnementaux. PhD Thesis, Université de Sfax and Université Bordeaux 1.

Ouellet, H. (1994). La recherche sociale et la politique de la santé et du bien-être. *Nouvelles pratiques sociales*, 7(1), 199–205.

Roselt/OSS CT7 (2004). Étude de la biodiversité dans l'observatoire pilote de Haddej – Bou Hedma (TUNISIE). Report, ROSELT/OSS,CT, Montpellier.

Roussiau, N. and Renard, E. (2003). Des représentations sociales à l'institutionnalisation de la mémoire sociale. *Connexions*, 80, 31–41.

Soini, K. and Dessein, J. (2016). Culture-sustainability relation: Towards a conceptual frame-work. *Sustainability*, 8, 167–178.

Souissi, M. (2008). Le tourisme dans les parcs nationaux en Tunisie. *Téoros*, 79–84.

Tarhouni, M., Ouled Belgacem, A., Neffati, M., Chaieb, M. (2007). Identification et caractérisation des systèmes écologiques dans le Parc National de Bou-Hedma. *Revue des régions arides*, 18, 25–43.

Zaafouri, M.S., Zouaghi, M., Akrimi, N., Jeder, H. (1997). La forêt steppe à *Acacia tortilis* subsp. *raddiana* var. *raddiana* de la Tunisie aride : dynamique et évolution. *Revue régions arides*, 258–271.

18

Architecture and the Biosphere Environment in Pedagogy: Design Visions for Sustainable Dwelling Communities

18.1. Introduction

Architecture today must evolve out of the bounds of its immediate built limits to extend into environmental surrounds and must concern itself more holistically with questions of natural preservation and sustainable development. Through design, architects need to inherently address questions related to the natural environment, and the integration of building and landscape, while promoting sustainable strategies that not only tackle the performance of the built structure but also extend to its natural context and community. In areas of biosphere reserves, especially in buffer zones, architectural development should involve conceptual, formal, behavioral and programmatic strategies that can support and connect to the extended community of the biosphere, through communal projects that integrate sustainable living, working and connecting to the natural terrain. This chapter will address the pedagogical experiments of engaging with the biosphere reserve of Jabal Moussa in Lebanon, through the work of 3rd-year architecture students at AUB. It will discuss the pedagogical methodology, the contextual framework of Jabal Moussa, and select design visions that try to imagine a potential sustainable housing community that can work symbiotically with the biosphere. As such, the work presented here reflects the necessity to address biosphere reserve areas through architecture in a holistic

---

For a color version of all the figures in this chapter, see www.iste.co.uk/romagny/biosphere2.zip.

Chapter written by Carla ARAMOUNY.

sustainable approach moving beyond the limits of the built form to encompass usage, integration, material and the extended natural and human environments.

18.2. Architecture and the environment

The work presented here starts from the premise that architecture today must work in synergy with nature and must perform symbiotically with it to enhance the performance of the built environment. Rather than reducing architecture to only its spatial and programmatic capacities, we need to rethink and redesign a more sensible and productive architecture that can be linked to its natural context and incorporate active environmental functions, thereby synergistically contributing to better both the built and natural environments. This is enabled by incorporating into the design process a deep understanding of the natural environment, its behaviors and environmental conditions, and establishing the possible synergies that can happen between the human habitat and the natural habitat in a nonintrusive sensible way. As such, and within this studio, the question of the natural environment in Lebanon is tackled through the particular focus on integration with biosphere reserves and their surrounding community through design, programming and environmental performance.

The concern for architecture's extended impact on the environment has been growing more urgently in recent years in the field, especially the larger impact of the built environment on our natural resources extending beyond the bounds of the building itself to the larger urban and territorial scales. Moving beyond the aspects of environmentally responsive solutions at the building scale, the intersections of the built environment with the natural environment at large need to be tackled from a more holistic ecological perspective, where both entities form a material, operational and social continuity. As such, the natural and built environments should be considered as forming a single and sensible ecology.

In his text "After Habitat, Environment", Steiner (2014) describes an evolution of the understanding of habitat, or the enfolding (natural or built) environment sustaining life, within contemporary architecture discourse. Referring to Conrad Waddington's work in the 1960s, he describes the latter's understanding of environment as a hybrid between technological/physical and natural interactions, between various ecologies and organisms.

This shift away from the language of habitat to that of environment, from regional territory and biology, to global informational networks, was thus marked by a loss of binary opposites as those between natural and social, open and closed systems, and city and country.

The shift towards recognizing the environment as a larger encompassing dynamic field intertwining both the man-made and the natural led, according to Steiner, to a habitat-based model of urbanism since the 1940s, brings forth the idea of "ecology" as model for conceiving and understanding the built environment in synergy with nature. Corner (2015) further elaborates on the ecological model in design as a dynamic organizational system that encompasses both landscape and architecture in flux and continuous feedback with one another.

The biosphere reserve in general presents an interesting condition of intersection between natural habitats and human habitats, between a natural reserve protecting particular ecosystems, and the human and communal habitats intersecting with it. The biosphere reserve as a condition thus brings forth this ecological model of co-habitat, where human, animal, built and natural environments work synergistically. In Steiner's essay, he refers to the work of evolutionary biologist Julian Huxely, who defined three ecological scales of interaction in habitats from the large climatic and regional scale, to the topological or terrain scale, to the biotic or immediate biological scale. The biosphere reserve involves these three ecological scales of habitat through its core, buffer and development areas, where interactions and feedback become crucial to sustain the life of the biosphere reserve.

Using this ecological model of understanding, our design studio approach was to work with nature and the biosphere of Jabal Moussa in particular in a more integrative manner. The projects developed by the students tackled the integration of new architectural interventions, specifically for housing communities in the buffer area of the biosphere, which could have a synergistic engagement and a more responsive attitude towards the natural environment.

18.3. Jabal Moussa Biosphere Reserve and the studio's premise

The studio focused on the Jabal Moussa Biosphere Reserve in Lebanon as an area of research and was developed in collaboration with the Nature Conservation Center at AUB, where architecture design and pedagogy became agents to engage students with biosphere reserves in research and conception.

Biospheres reserves in general are protected extended environments that integrate natural, social and cultural heritage, while supporting adjacent communities and people. Under the program Man and Biosphere developed by the UNESCO, the Jabal Moussa Biosphere Reserve was granted its status as a preserved area in 2009 (see the Jabal Moussa website: https://www.jabalmoussa.org/reserve).

It is a mountainous area surrounded by seven villages between the Keserwan and Jbeil districts, with a rich natural biodiversity from flora and fauna, to cultural heritage and ruins dating back to Roman and Phoenician periods. The reserve further incorporates eco-tourism through its hiking trails, guesthouses, in addition to local produce markets and a tree nursery. Through its different aspects, the reserve relies on and supports the adjacent local communities, from villagers in the region who work in it, to the establishment of guesthouses, small eateries and villagers' kitchen. The Jabal Moussa Biosphere Reserve is formed out of three zones, the core natural zone, the buffer area including towns such as Qehmez and Mchati on its southern edge and the larger development or transitional area including further villages and towns.

The studio recognized the Jabal Moussa Biosphere Reserve as a valuable zone that enables regional development, and that has the potential to integrate new programs, especially in its buffer areas through an engagement between the central natural core and the outer development and peri-urban areas. The students were asked to imagine architectural interventions in the buffer areas around the reserve, to incorporate viable housing strategies that reshape the possibilities for sustainable domesticity. The idea was to incorporate new communal models for living and working, close to these natural zones, and away from the polluted urban areas, allowing inhabitants to escape from the mundanity of suburban housing into a communal living that reconnects them back to nature. Furthermore, these eat–live–work communities would support the reserve and enhance the connection between peripheral urban areas and the close-by villages. The preferred sites for such interventions were selected in areas within the buffer zone that are within proximity to dense urban environments, as the project aims to propose viable autonomous communities that can serve as the link between the biosphere reserve and suburban centers. The chosen sites included upper and lower parts of the village of Qehmez and the village of Mchati near the Nahr el Dahab river.

18.3.1. *Studio methodology and research in three scales of operation*

Based on the previously mentioned ecological model, the studio methodology was developed through three main scales of research, understanding and design operations:

– The biosphere reserve scale/engage: the first scale is that of the entire biosphere and includes a proactive engagement with it through rigorously understanding its conditions and learning about its mechanisms and the potential for architecture and people to contribute further to its livelihood.

– The habitat scale/perform: this involves the scale of the habitat and its integration with the immediate contextual and climatic conditions of the site for the housing project, by enabling the architectural space to react, behave and operate in symbiosis with the outside environment.

– The human/organism scale/activate: the third scale is that of the community, where the research focused on enabling new socio-cultural user interactions, communal activities and supporting programs that benefit and synergistically connect with the towns and the biosphere environment.

18.3.1.1. *The biosphere reserve scale: engage through a proactive approach*

This first scale involved deep understanding of biosphere reserves in general and focused on the Jabal Moussa context and its surrounding villages of Mchati, Qehmez Plateau (upper) and Qehmez Valley (lower). Through in-depth research, mappings and analysis, the students gained thorough knowledge of the context within which their architecture projects could be imagined. Their investigations of the area were approached from different perspectives but focused generally on climatic and environmental factors, local ecosystems, spatial typologies and socio-economic realities.

The research used available resources from books, references, and online information and maps. However, to understand the site more intimately, a hiking and weekend trip was organized with the Nature Conservation center's support for the students and professors to engage directly with the biosphere. The group visited the biosphere area and explored it through the different walking trails, and spent time in the local nearby towns and their available facilities from the eating houses to local small markets. They also interacted with the local community to help formulate a clearer picture about the intersections of Jabal Moussa with the community surrounding it.

The students' research tackled different aspects of biosphere reserves in general and the Jabal Moussa Biosphere Reserve in particular. They understood biosphere reserves first as the rich ecosystem and life-sustaining layer on Earth that forms the habitat of all living creatures, living together in symbiosis. The UNESCO Man and Biosphere program moves from this premise to identify and protect rich biospheres around the world that designate particularly rich biodiversity, community integration and cultural and natural heritage, while enabling economic and sustainable development. Each biosphere is structured along three interrelated areas: the core, buffer and transition zones. As one of the 33 UNESCO Biosphere Reserves in the

Arab World, Jabal Moussa is understood as a particularly rich landscape and community, home to protected species and local cultural heritage, and engaged with 7 surrounding villages. It is a mountainous area in the Keserwan district of Lebanon (Figure 18.1), bound by two rivers and covering an area of 6,500 hectares, at an altitude ranging between 350 meters and 1,700 meters (MAB Med 2015). It includes a rich ecosystem ranging from riparian in the valleys to open woodland in the mountainous area, with varied and unique flora and fauna. In its forest area, it includes species such as Oak, Juniper, Pine and wild Apple trees, in addition to various types of wild orchids and endemic plant species such as the cyclamen and the Lebanese oregano. Jabal Moussa is also an important migratory bird habitat, and a bottleneck for migratory routes, with bird species such as eagles, storks and sparrow hawks (Figure 18.2). Different species of wild animals also inhabit Jabal Moussa, such as the striped hyena and wild boar, with some particular to the Middle East region and Lebanon like the rock hyrax (Figure 18.3). The students in their research further understood different climatic conditions in the area, from wind direction and speed in different seasons, sun orientation and varying humidity levels (Figure 18.4). Additionally, they looked at socio-economic and bio-integrated programs that form part of the richness of Jabal Moussa, from recreational eco-tourism activities to medicinal plants, honey making and local food produce (Figure 18.5).

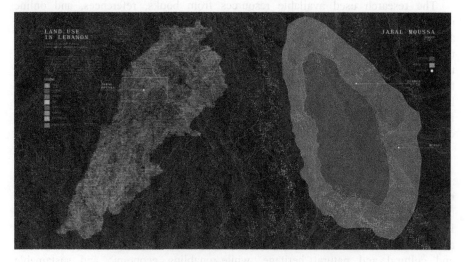

Figure 18.1. *Jabal Moussa Biosphere Reserve in Lebanon (Lea Tabaja, Yara Haidar, Noura Bissat)*

Figure 18.2. *Flora, fauna and migratory birds (Lea Tabaja, Yara Haidar, Noura Bissat)*

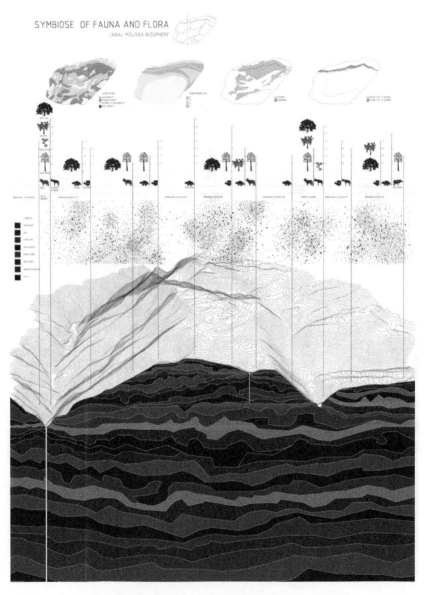

Figure 18.3. *Wild animals of Jabal Moussa (Myriam Abou Adal, Marc Faysal, Amir Moujaes)*

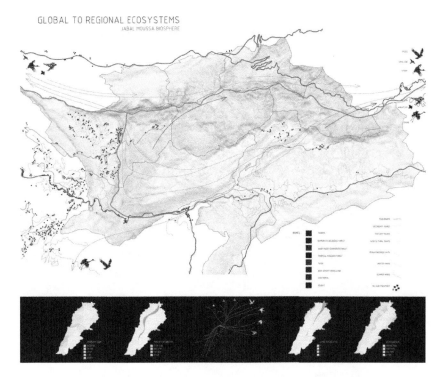

Figure 18.4. *Wind speed, seasonal flows and local villages (Myriam Abou Adal, Marc Faysal, Amir Moujaes)*

Furthermore, they zoomed in to select areas around the villages of Mchati at Nahr el Dahab and Qehmez (upper and lower sections of the town), to understand in more depth their various environmental and social characteristics. The upper plateau of Qehmez was understood as a productive agricultural plain with links to Jabal Moussa through a main entrance into the reserve, and proximity to local guesthouses and the tree nursery (Figure 18.6). The lower section of Qehmez in the valley was also understood as a rich agricultural area with tomato plantations and affected by a direct proximity to local uncontrolled quarries (Figure 18.7). The area featured traditional single-unit house typologies, with terraces and vine shaders in connection to the agricultural fields. The village of Mchati was also researched and understood as a town that is adjacent to the Nahr el Dahab river. Featuring a series of terraced zones on the side directly adjacent to the steep slopes of the biosphere, Mchati includes many direct supporting programs to Jabal Moussa, from a local produce market to a guesthouse and an eatery that engage with visitors of the biosphere (Figure 18.8).

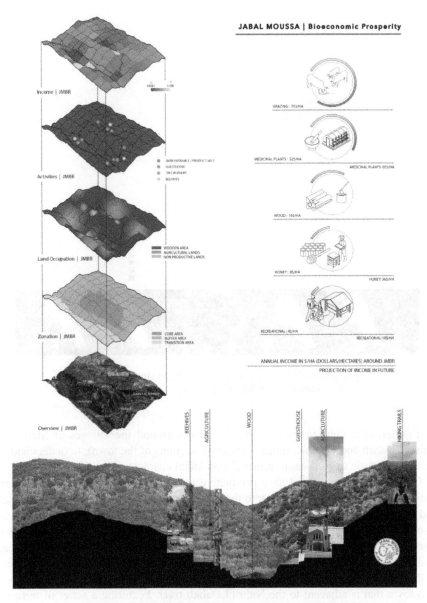

Figure 18.5. *Local activities and bioeconomic potential (Joseph Chalhoub, Aya el Husseini, Baraa Al Ali, Samer Abboud)*

Architecture and the Biosphere Environment in Pedagogy 275

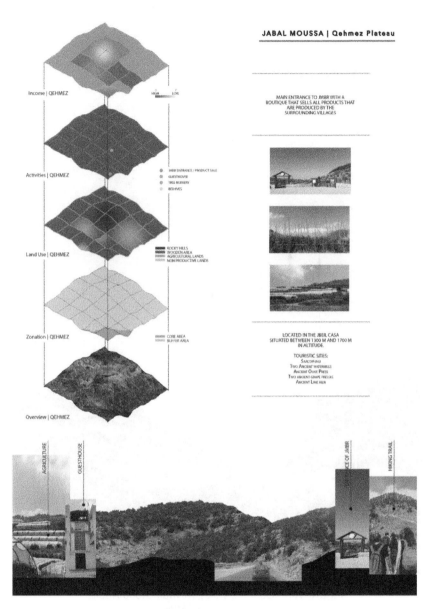

Figure 18.6. *Qehmez plateau (Joseph Chalhoub, Aya el Husseini, Baraa Al Ali, Samer Abboud)*

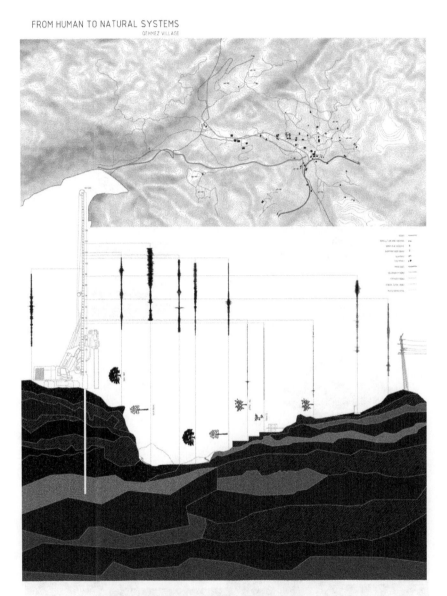

Figure 18.7. *Qehmez valley (Myriam Abou Adal, Marc Faysal, Amir Moujaes)*

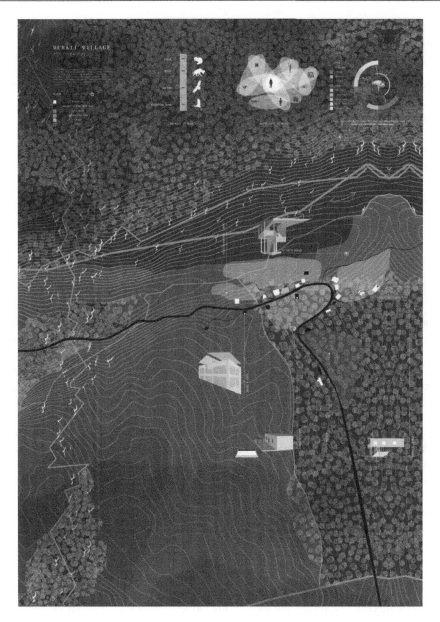

Figure 18.8. *Village of Mchati (Lea Tabaja, Yara Haidar, Noura Bissat)*

18.3.1.2. *The habitat scale: perform through environmental responsiveness and housing*

The second scale of the studio process involved understanding environmental behavior and site conditions as extensions of an architectural space through a design process of abstract physical models. The environment and climatic conditioning of an architectural building were looked at in depth within this studio. Inspired by works of architects like Philip Rahm, the approach involved understanding climatic and atmospheric conditions, from heat, wind, humidity and others, as dynamic flows that are implicated by the form and spatial parameters of an architecture project.

> The aim is to conceive an architecture free of any formal and functional predetermination: variable, fluctuating, open to meteorological permutations and the passage of time, to seasonal changes, to the alternation of night and day and moreover to the sudden appearance of unanticipated functions and forms (Rahm 2007).

To start the conceiving of their model experiments in relation to climatic conditions, the students narrowed into a site of their choice in one of the three selected villages. They began their conceiving of an architectural spatial intervention that can foster good climatic and sustainable synergy with its site and context. The students understood through further research issues such the site's materials and landscapes, its orientation, as well as wind and energy flows, sustainable materials, passive design strategies and typological integration with landscape.

The design process used to achieve this focused on developing physical experimental models at a scale of 1/20. The models were developed as abstract spaces that intersect environmental behaviors and performance with architectural typologies. Drawing on specific site conditions from wind direction, sun orientation and others, the students each created and imagined a space that optimizes climatic behavior while creating an interesting architectural and landscape experience. Accordingly, each produced model intersected three main parameters:

– Natural elements: the space in the model should help enhance or reduce the behavior and flows of natural climatic elements, such as heat, cold, air and wind circulation, rain and humidity.

– Architectural typology: the form of the model should start from specific spatial typologies (such as courtyard, tower, elevated mass or others) that can optimize or transfer these natural flows. The selected typologies should also provide a meaningful spatial experience

– context: the architectural model should also articulate or respond to a condition of the site, such as a tree or landscape species, a water feature, a rock formation or another locally found characteristic.

Each of the students' models thus provided a climatic concept and formal approach that allowed them later to design a more developed housing scheme. Moving from these abstract experiments into housing typologies, each of the houses designed by the students became thus a spatial result that intersects with site issues, landscape and climatic behaviors. This abstract experimental phase allowed the students to understand architectural space and form a parameter that controls climatic energy flows and enables certain levels of comfort and shelter. It also allowed the students to factor in spatial experience and living atmosphere while critically solving climatic needs (Figures 18.9–18.11).

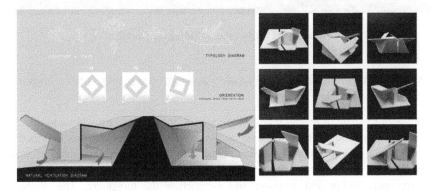

Figure 18.9. *Climatic model (Joseph Chalhoub)*

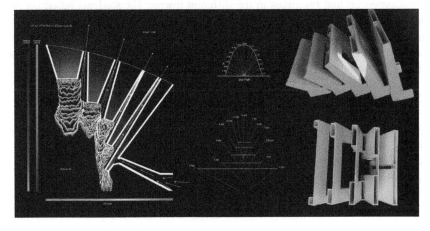

Figure 18.10. *Climatic model (Omar Ayache)*

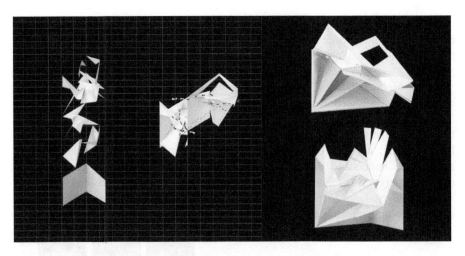

Figure 18.11. *Climatic model (Lea Tabaja)*

18.3.1.3. *The human/organism scale: activate programs and community engagement*

In the third scale of research within the studio, the students moved on to develop and propose potential programs and communities, which could be imagined to relocate to the vicinity of Jabal Moussa. Based on their in-depth site research, each of the students delved into their own programmatic proposal, foreseeing a new type of community living and working and supporting the biosphere. The proposed program served as a link between the incoming inhabitants and the local area and its people, and as a potential communal and sustainable space that can link to the existing synergies and resources in the different towns. The students relied on their investigations of local services, the different towns, their resources and produce, in addition to currently existing synergies between the biosphere reserve and the villages, to propose new types of programs that can be situated in the proposed housing community and that can engage with the biosphere reserve and villages.

For example, one of the students had looked at existing economic potentials in the village of Mchati and found that honey making can be a good source of production due to precedents in the area. The student's new programmatic proposal centered around beekeeping and honey production through a community including beekeepers and researchers, living together and co-producing honey to be sold through the biosphere reserve's shops and markets (Figure 18.12).

Architecture and the Biosphere Environment in Pedagogy 281

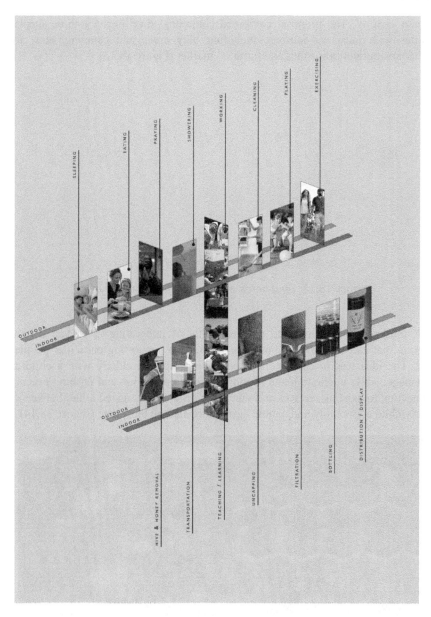

Figure 18.12. *Honey-making community program (Joseph Chalhoub)*

Another example looked at a program on a site near Qehmez that can seasonally house local nomad shepherds, who move yearly between the Bekaa valley and Jabal

Moussa vicinity. The proposal combined communal dwellings for three shepherd families with local eco-tourism, including play areas and camping sites, food production, agricultural zones and farmers' market (Figure 18.13).

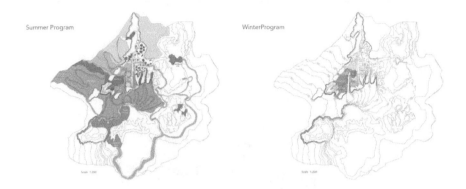

Figure 18.13. *Shepherds and eco-tourism (Omar Ayache)*

A third example considered the possibility of integrating in Mchati a hybrid program that can support the biosphere, while also supporting particular people in need. The idea was to combine a nursing home for the elderly with a children's orphanage, while integrating and supporting food production and farming practices. Connecting to existing terraces and vines, the program instigated a link between the elderly/child community and nature, agriculture and the biosphere (Figure 18.14).

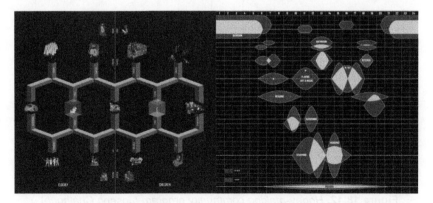

Figure 18.14. *Housing for the elderly and children (Lea Tabaja)*

18.3.2. *Student architectural proposals and results*

Following up on these different scalar investigations, from site research, climatic model experiments, to programmatic proposals, the students moved then to develop their architectural proposal for the dwellings, by first starting with an individual housing unit to accommodate a single individual or a family. The design of each of the housing units relied on the programmatic needs of their proposed program and users, in addition to climatic behavior, spatial experience and internal connectivity. Each design of the house had to establish a clear relationship to the outdoors and the natural site, and to involve a sensitive integration with the landscape and local materials. From the design of the unit, the students then moved to develop strategies to combine the individual units together to form communal clusters and agglomerations, and to conceive of their overall community scheme including additional production and recreational areas. Within the communal agglomeration, the houses needed to interact together both spatially and socially, as they respond to site, climatic and programmatic conditions.

Different student projects developed out of the semester, each approaching the three scales of research uniquely, and resulting in varied design visions that cross communal housing with environmental integration and biosphere synergies.

The first project focused on the socio-economic potential of the area, and in the village of Mchati specifically, through incorporating a productive community interacting with the biosphere through honey making. As developed by student Joseph Chalhoub, the project for the community housing and honey production facilities is developed as an integrated architecture within existing terraces on the Mchati slopes, facing the river and adjacent to Jabal Moussa's lower entrance. In his proposal, Joseph focused on designing first a single housing unit that sits within the slope and that relies on his earlier physical models to design large southern openings and ventilation flows that optimize the climatic conditions of the unit. To create the community, he went on to cluster each of the two houses together, centered around a shared outdoor landscape, with all of the other clusters aligning to different levels of the existing terraces. A central communal outdoor area links the inhabitants of this community together. The community would maintain beehives in close-by lots adjacent to the biosphere and use their communal and production spaces at the lowest end of the development, to produce honey as an important local produce. The design also makes use of local stone materials and local plants and agricultural to create a holistic and productive environment (Figure 18.15).

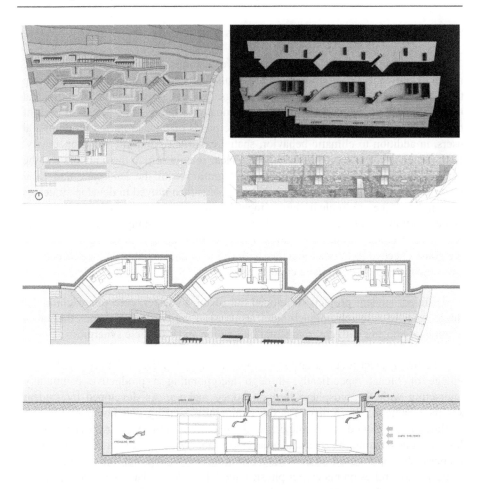

Figure 18.15. *Integrated community dwellings and honey production (Joseph Chalhoub)*

A second project by Omar Ayache focused on embedding architecture within the natural landscape and on supporting existing vulnerable communities in the vicinity of Jabal Moussa. Omar started by working on an existing rocky area in Qehmez, with varied geological formations. In his first part investigating humidity and the resulting formation of underground cavities through erosion, Omar experimented with different relationships between heat and humidity transfer, and conditions of height and width of subterranean spaces. His intervention involved casting and creating clay layers with vertical inlets within the artificial rock, allowing it to erode and develop spaces and cavities in time with varying heat and humidity levels. This new rock formation would merge together with the site's natural limestone and form well-conditioned spaces over time. His previously mentioned program that crossed housing and shelter needs for local shepherds also included local and visitor programs, from a farmer's market, to guesthouses and camping sites. The program was situated within the architectural cave-like space following geothermal factors and comfort needs, ranging from higher warmth to lower humidity levels. By mainly housing three families of shepherds, the community would thus revolve around the interaction with the shepherds, the local farmers and local tourists to the biosphere and the area. The design itself is formed out of the natural process of erosion and is created as a multileveled house, where each room is positioned according to heat and coolness factors. The houses also open up to and connect to one another through communal areas and meeting zones (Figure 18.16).

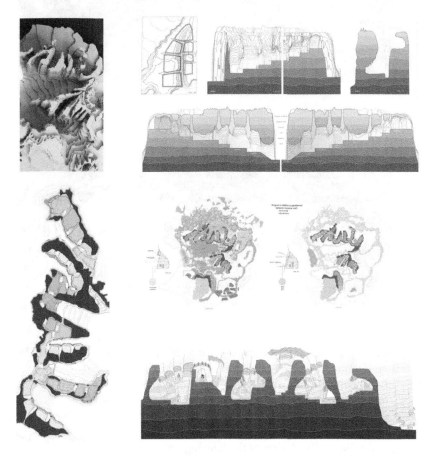

Figure 18.16. *Embedded geological dwellings for local nomads (Omar Ayache)*

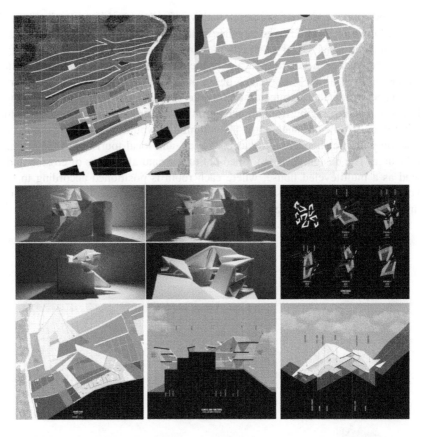

Figure 18.17. *Hybrid architecture and existing landscape terracing (Lea Tabaja)*

The third project by Lea Tabaja focused on evolving a new hybrid community integrated on the agricultural terraces of Mchati, specifically on a zone with existing vine climbers. Her project began through complex studies of intersecting surfaces on a slope, which could optimize cross-ventilation and heat transfer. The existing vines on the site created an underlying grid that guided Lea's decision on where to intervene. Her program proposed that a group of elderly would move to and live in a housing community that also includes a children's home. The two groups, the elderly and children, would live within the same community, sharing intersecting spaces, from eating areas to gathering zones, while also having separate functions such as private bedrooms, reading rooms and playing rooms. They would thus inhabit together this communal dwelling while engaging with the adjacent biosphere and the local village. The design of the housing units started from the earlier formal and climatic experiments, and then evolved to incorporate programmatic needs and slope requirements. Each unit is designed as part of a dual cluster, with two homes engaging together with different types of supporting spaces. The programs were positioned relative to individual and common functions, and also based on the users' day/night activities and comfort needs. Sleep spaces, for example, were located on similar private sides within the cluster, while more active play or activity areas were located in more open and connected spaces to the outdoors. The existing vines on the slopes further enabled an outdoor/indoor productive connection to landscape (Figure 18.17).

Figure 18.18. *Productive housing clusters and the environment (Myriam Abou Adal)*

The fourth project by Myriam Abou Adal focused on an environmentally responsive architecture and proposed that a group of botanists would relocate to Qehmez to experiment and research on the different species in the Jabal Moussa area. The botanists would live together in this sustainable community, where the work and living programs intersect and meet. Different outdoor agricultural areas and indoor greenhouses form part of the community and engage directly with the houses and lab spaces. The design of the housing units, and then later the clusters, were focused on an optimized indoor climatic performance with protection from harsh sun rays, while enabling thermal heat absorption by the walls and planted roof. Good cross-ventilation and vertical stacking of spaces also allowed a continuous hot airflow throughout the space. The planted roofs further served as outdoor meeting areas and productive landscapes that serve the users. From a macro perspective, the entire community was designed to engage with the rocky site of Qehmez, while following the constraints of the slope, and intersecting architecture with landscape and infrastructural needs. The common areas included outdoor agricultural zones, a walking/exercising track connecting all of the units and a water collection system that provided for water needs in summer. Common supporting programs also included waste composting and a seed bank that would serve as an educational facility to incoming visitors (Figure 18.18).

18.4. Conclusion

The studio thus worked on bringing research and understanding of biosphere reserves right into the heart of the architectural studio, and to position it as an essential subject that students could engage with to propose more holistic and sensitive integrations of integrate architecture with environment. The biosphere itself served as a model of reference, as an ecology that includes necessary synergies between people, the natural environment, in all of its encompassing richness from the ground layers to the atmosphere. Architecture as such was seen and developed as a continuity of this ecosystem, and as an inherent synergistic space that engages with its surrounding and community. Through a scalar research methodology that moves from the scale of the biosphere, the habitat, to that of the community, the work developed in this studio allowed students to design novel intersections of architecture and the environment, and to propose new visions for housing communities.

Using research, visualization and experimental climatic models, the students were able to develop their projects as potential alternatives to urban or rural housing typologies, and to cross sustainable design strategies with productive landscapes. The aim overall was to develop new architectural ideas that can be situated in the buffer areas of Jabal Moussa and that could become crucial supporting programs to the life and development of the biosphere.

Course Instructors: Carla Aramouny, Nicolas Fayad, Sandra Frem.

18.5. References

Banham, R. (1984). *The Architecture of the Well-Tempered Environment*. The University of Chicago Press, Chicago.

Belanger, P. (2015). *Going Live: From States to Systems*. Princeton Architectural Press, Princeton.

Corner, J. (2015). Organizational ecologies. In *Going Live: From States to Systems*, Belanger, P. (ed.). Princeton Architectural Press, Princeton.

De Lapuerta, J.M. (2017). *Collective Housing: A Manual*. Actar Publishers, New York.

Gomez Luque, M. and Jafari, G. (2018). *New Geographies 9: Posthuman*. Harvard University Graduate School of Design and Actar, Cambridge.

Ibañez, D. and Katsikis, N. (2014). *New Geographies 6: Grounding Metabolism*. Harvard University Graduate School of Design, Cambridge.

MAB Med (2015). A walk through Jabal Moussa [Online]. Available at: https://www.jabalmoussa.org/sites/all/themes/jabalmoussa/img/walk-through.pdf.

Matar, D. (ed.) (2010). *Jabal Moussa: Between Myth and Reality*. Association for the Protection of Jabal Moussa, Kesrouan.

Mostafavi, M. (2010). *Ecological Urbanism*. Lars Müller Publishers, Baden.

Rahm, P. (2007). Form and function follow climate. *AA Files*, 2–11 [Online]. Available at: https://www.jstor.org/stable/29544645.

Rahm, P. (2019). *Architectural Climates*. Lars Müller Publishers, Baden.

Steiner, H. (2014). After habitat, environment. In *New Geographies 6: Grounding Metabolism*, Ibañez, D. and Katsikis, N. (eds). Harvard University Graduate School of Design, Cambridge.

List of Authors

Sonia ADERGHAL
Mohammed V University
Rabat
Morocco

Abdelaziz AFKER
Water and forest engineer
Salé
Morocco

Lahoucine AMZIL
Mohammed V University
Rabat
Morocco

Carla ARAMOUNY
Saint-Joseph University
Beirut
Lebanon

Hicham ATTOUCH
Mohammed V University
Rabat
Morocco

Lahcen AZOUGARH
Ibn Tofail University
Kénitra
Morocco

Joelle BARAKAT
Jabal Moussa Biosphere Reserve
Lebanon

Angela BARTHES
University of Aix-Marseille
Digne
France

Martí BOADA
Autonomous University of Barcelona
Spain

Antonio BONTEMPI
Autonomous University of Barcelona
Spain

Saïd BOUJROUF
Cadi-Ayyad University
Marrakech
Morocco

Soukaina BOUZIANI
Mohammed V University
Rabat
Morocco

Abdelkarim BRAHMI
University of Gafsa
Tunisia

Driss CHAHHOU
Mohammed V University
Rabat
Morocco

Abdelkader CHAHLAOUI
Moulay Ismail University
Meknes
Morocco

Catherine CIBIEN
MAB Programme
UNESCO
Paris
France

Tarik DAHOU
Institut de recherche pour le développement (IRD)
Paris
France

Mchich DERAK
DREFLCD
Rif
Morocco

Pierre DOUMET
Jabal Moussa Biosphere Reserve
Lebanon

El Habib EL AZZOUZI
Mohammed V University
Rabat
Morocco

Hind EL BOUZAIDI
Mohammed V University
Rabat
Morocco

Yamina EL KIRAT EL ALLAME
Mohammed V University
Rabat
Morocco

Faiza EL MEJJAD
Cadi-Ayyad University
Marrakech
Morocco

Mohammed FAEKHAOUI
Mohammed V University
Rabat
Morocco

Fatimazahra HAFIANE
Mohammed V University
Rabat
Morocco

Kawtar JABER
Saint-Joseph University
Beirut
Lebanon

Maya KOUZAIHA
Saint-Joseph University
Beirut
Lebanon

Nadia MACHOURI
Mohammed V University
Rabat
Morocco

Roser MANEJA
Autonomous University of Barcelona
Spain

Alejandro MARCOS-VALLS
Autonomous University of Barcelona
Spain

Antoni MAS-PONCE
Autonomous University of Barcelona
Spain

Wahiba MOUBCHIR
École Normale Supérieure
Marrakech
Morocco

Ahmed MOUHYIDDINE
Mohammed V University
Rabat
Morocco

Reda NACER
Mohammed V University
Rabat
Morocco

Ken REYNA
Regional Natural Park of
Mont Ventoux
Carpentras
France

Bruno ROMAGNY
Institut de recherche pour le
développement (IRD)
Marseille
France

Maria Carmen ROMERA-PUGA
Autonomous University of Barcelona
Spain

Rachid SAMMOUDI
Mohammed V University
Rabat
Morocco

Sònia SÀNCHEZ-MATEO
Autonomous University of Barcelona
Spain

Index

A, B

agriculture, 10, 21, 24–26, 32, 48, 50, 81, 85, 87, 89, 96, 99, 116–118, 126, 128–130, 143, 144, 153, 169, 177, 178, 185, 200, 205, 210, 228–230, 245, 257, 273, 282, 283, 286, 288

architecture (*see also* habitat), 254, 265–269, 278, 279, 283, 284, 286–288

argan, 105, 107, 110, 115, 117–119, 128, 130–132

arganeraie, 12, 23–25, 105–113, 118, 119, 123, 126–129, 132, 134, 136, 138

biodiversity, 5–10, 12, 18, 20, 23, 26, 28, 36, 41–43, 45, 55, 74, 77, 78, 80–82, 85–89, 100, 102, 107, 108, 110, 112, 114, 117, 119, 123, 124, 126, 127, 133, 142, 143, 145, 146, 165, 167, 170, 184, 187, 192–195, 205, 214, 220, 241, 247, 251, 252, 256, 259, 261, 268, 269

C

climate change, 45, 52, 89, 96, 114, 205, 209, 211–214

community(-ies), 6, 15, 16, 22, 58, 102, 112, 128, 131, 141, 143, 146, 165, 197, 206, 265–269, 280–284, 286, 288
local, 268, 269

conservation, 5, 7, 11, 13, 15, 17, 18, 23, 24, 26, 27, 33, 34, 41, 45, 49, 52, 54, 55, 60, 62, 64, 77, 81, 82, 88, 98, 102, 103, 107, 108, 110, 113–115, 117, 119, 120, 124, 126, 129, 132, 133, 136–138, 154, 161, 165, 170, 173, 175, 177, 184, 219–228, 230, 232, 233, 238, 242, 244–246, 256, 258, 267, 269

F, H

fires, 116, 165–170, 176, 189, 191, 193, 223, 225, 228

fishing (*see also* trawling), 24, 31, 32, 48, 85, 126, 225, 230, 231, 233–247
maritime, 237, 238, 246

France, 6, 8, 9, 11, 12, 14, 18, 19, 21, 22, 27, 31, 32, 39, 96, 167, 235, 238

habitat (*see also* architecture), 53, 59, 60, 107, 143, 234, 266, 267, 269, 278, 288

heritage, 6, 8, 10, 20, 26, 34, 36, 41, 44, 45, 64, 81, 82, 88, 100, 117, 118, 123, 124, 128–133, 138, 141, 143, 144, 166, 255, 256, 258–261, 267, 269

K, L

knowledge, 115, 128, 146, 220, 232, 247
  local, 115, 232
  scientific, 247
Lebanon, 6, 7, 9, 12, 25, 95, 96, 98, 102, 202, 265–267, 270

M, O

management/administration, 5–7, 10–14, 16–18, 20, 21, 24–26, 28, 31–34, 36, 40–42, 44, 52, 59, 60, 64, 69, 78, 79, 83, 85, 87, 88, 99–101, 105, 110–112, 115–117, 119, 120, 124, 127, 129, 130, 132–138, 141, 143–146, 167, 170, 173, 194, 197, 214, 215, 220, 223, 224, 226, 228, 230–233, 236–238, 246, 258, 260
  plan, 20, 24, 34, 102, 220, 226–228, 231–233, 237, 247
Morocco, 6, 7, 9, 12, 23, 27, 46, 47, 105, 113, 114, 123, 124, 126, 127, 129, 130, 141–143, 145, 167, 169–171, 174, 178, 183, 185–189, 192, 193, 195, 196, 202, 205, 209, 210, 212, 213
oasis, 23, 25, 141–145

P, R

patrimonialization, 6, 115, 124, 128, 130–133, 135, 136, 138, 139
pedagogy, 267
pesticides (*see also* residues), 153, 154, 156–161, 189, 229
protected areas, 7–9, 14, 18, 20, 23, 26, 27, 32, 41, 46, 99, 108, 161, 195, 219, 233, 251, 262
residues (*see also* pesticides), 153, 154, 158, 160, 161

S, T

social
  and solidarity economy, 173–175
  representations, 251–256, 261, 262
Spain, 6, 8–11, 14, 26, 27, 42, 43, 46, 74, 79, 167, 178
sustainability, 27, 44, 45, 69, 73, 76, 78, 79, 81, 82, 86, 88, 89, 115, 124, 143, 145, 174, 199, 238, 240, 244, 265, 268, 269, 278, 280, 288
sustainable development, 5, 7–9, 11, 13, 14, 44, 45, 69, 73, 74, 77–79, 81, 82, 87–89, 98, 110, 112, 115, 118, 119, 123, 126, 127, 130, 137, 138, 142–144, 161, 165, 169, 178, 184, 187, 195, 210, 220, 254, 255, 258–260, 262, 263, 265, 269
trawling (*see also* fishing), 240, 242, 245
Tunisia, 237, 251, 252, 256–261

Summary of Volume 1

Presentation of the Authors of the Two Volumes

Introduction
Angela BARTHES, Catherine CIBIEN and Bruno ROMAGNY

Part 1. Biosphere Reserves and Sustainable Development Goals: Multidisciplinary Scientific Issues

Introduction to Part 1
Bruno ROMAGNY

Chapter 1. Man and the Biosphere: A Precursory Program for the Next World
Meriem BOUAMRANE and Didier BABIN

    1.1. 1971–2021, the beginnings of sustainable development
    1.2. Making sure no one is left behind
    1.3. Identification of gaps, risks and challenges
    1.4. Valuable lessons learned from the transformation towards sustainable and resilient societies
    1.5. Investments that may affect the building of sustainable and resilient societies
    1.6. Integration of biodiversity within sustainable development policies
    1.7. Policy recommendations to accelerate progress in building sustainable and resilient societies
    1.8. Lessons learned from the Covid-19 crisis and perspectives for biosphere reserves for the next world
    1.9. References

Chapter 2. Humans and Nature: A Story to be Rewritten
Magda BOU DAGHER KHARRAT, Éliane BOU DAGHER and Rhéa KAHALÉ

    2.1. *Homo sapiens*, a species like the others
    2.2. *Homo sapiens*, a nature modifier
    2.3. The Mediterranean, more than a sea in the middle of the land
    2.4. The academic sphere and the action in favor of biodiversity
    2.5. Biosphere reserves and Sustainable Development Goals
    2.6. References

Chapter 3. Social Representations, Collective Organization and Mediterranean Biosphere Reserves
Angela BARTHES, Bruno ROMAGNY, Jean-Marc LANGE, Lahoucine AMZIL, Roser MANEJA, Mohammed ADERGHAL and Véronique CHALANDO

    3.1. Introduction
    3.2. Social representations as an exploratory method of prior knowledge
    3.3. How can social representations be defined? Some theoretical elements
    3.4. How can social representations be defined? Central core and peripheral elements
    3.5. The methodological elements of our study
    3.6. Study results
    3.7. Differences and similarities in the social representations of students
        3.7.1. Social representations relatively shared by students enrolled in France and in Spain
        3.7.2. Notably different results for the students enrolled in Morocco
    3.8. Addressing the issue of complexity versus focusing on the environment
    3.9. Addressing the collective organization of society versus the recourse to individual action
    3.10. Conclusion
    3.11. References

Chapter 4. Challenges and Opportunities of Collaborative Research on Biosphere Reserves in the Mediterranean
Moustapha ITANI, Salma NASHABE TALHOUK, Wassim EL-HAJJ, Nivine NASRALLAH and Hannah ABOU FAKHER

    4.1. Introduction
    4.2. Collaborative research
    4.3. Beneficial aspects of collaborative research
    4.4. Challenges to collaborative research and data sharing
    4.5. Motives behind collaborative research
        4.5.1. Components of collaborative research
        4.5.2. External components
        4.5.3. Internal components

4.6. The Mediterranean Basin: asymmetries between Northern and Southern Mediterranean countries
    4.6.1. Economic development
    4.6.2. Human and social development
    4.6.3. Trade and economic integration
    4.6.4. Scientific contributions and representation
    4.6.5. Impediments to collaboration across the Mediterranean
    4.6.6. Regionalism and conflict
    4.6.7. Academic boycotts
4.7. Travel limitations
    4.7.1. Language barriers
    4.7.2. Institutional structures promoting collaborative research in the Mediterranean
4.8. Conclusion
4.9. References

Chapter 5. Scientific Tourism in Multi-Labeled Protected Areas: The Ecological Transition and Controversy in the Mountains
Mikaël CHAMBRU and Cécilia CLAEYS

    5.1. Introduction
    5.2. The ecological transition: from the injunctions to the different socio-political and cultural references
    5.3. The trajectories of governance forms for a scientific tourism project
    5.4. The ambiguities related to the touristic development of scientific culture
    5.5. The environmental paradoxes of a scientific tourism project
    5.6. Conclusion
    5.7. References

Part 2. Educational Practices Relating to Biosphere Reserves: Balance and Prospects

Introduction to Part 2
Angela BARTHES

Chapter 6. Teaching How to Produce Differently at a Biosphere Reserve
Véronique CHALANDO and Angela BARTHES

    6.1. Introduction
    6.2. Curricular challenges of teaching how to "produce differently"
        6.2.1. Curriculum development model
        6.2.2. Distinction between agroecology/agro-ecology: the question of the reference materials
    6.3. Technical knowledge and political movements

6.4. Knowledge conflicts and conflicts of values: the question of direction in the circulation of knowledge
6.5. Towards coherent criteria for analyzing agroecological literacy
    6.5.1. Essential curriculum elements
    6.5.2. A practical method for measuring necessary curricular links
6.6. Case study
    6.6.1. Biosphere reserves, levers to "produce differently"?
    6.6.2. The Mont Ventoux biosphere reserve and the agricultural school of Carpentras
    6.6.3. Case study results
6.7. Discussion
    6.7.1. What curricular coherence should exist for teaching to "produce differently" in the biosphere reserve?
    6.7.2. The teacher's posture in relation to agroecology
    6.7.3. The role of the territory in the implementation of a consistent curriculum
6.8. Conclusion
6.9. References

Chapter 7. The Sustainable Management of Biosphere Reserves: What Are the Challenges for Agricultural Education?
Nina ASLOUM, Guillaume GILLET and Laurent BEDOUSSAC

7.1. Introduction
7.2. Agroecology, from its emergence to the change of agricultural model
    7.2.1. Agroecology: an ever-evolving, polysemous concept
    7.2.2. Agroecology for a real change in the agricultural model?
7.3. Social representations
    7.3.1. The theoretical framework of social representations
    7.3.2. The structural approach to social representations
7.4. Methodology
    7.4.1. Data collection
    7.4.2. Populations
7.5. Data categorization
7.6. Results
7.7. Discussion
7.8. Conclusion
7.9. References

Chapter 8. Collective Skills from Partnerships Between Protected Areas and Teachers
Sylviane BLANC-MAXIMIN

8.1. Introduction

8.2. The educational partnership
    8.2.1. Partnerships in national education
    8.2.2. A partnership for creating collective professional skills?
8.3. Three case studies in a labeled rural territory
    8.3.1. Descriptions of the projects undertaken in the three case studies
    8.3.2. Data collection methods
    8.3.3. Results
8.4. Presence of a collective skill and of the collective's skill
    8.4.1. Case A in school one
    8.4.2. Case A in school two
    8.4.3. Case A in school three
    8.4.4. Case B
    8.4.5. Case C
    8.4.6. Synthesis
8.5. Conclusion
8.6. Appendix
8.7. References

Chapter 9. The Instrumentalization of Education in Sustainable Development at the Service of Tourism: The Case of the Arganeraie

Salma ITSMAÏL and Bruno GARNIER

9.1. Introduction
9.2. Environmental crisis and inflation of alternative tourism
9.3. Tourism and sustainable development
9.4. Sustainable tourism and patrimony: educational issues
9.5. Towards a "sustainable strategy"
9.6. The Moroccan situation: a sustainable tourism policy in the ABR?
9.7. A cultural as well as a natural patrimony item: the argan tree
9.8. Between reality and opportunism: the instrumentalization of sustainable development
9.9. Education: the missing vector for sustainable tourism
9.10. Conclusion
9.11. References

Chapter 10. Biosphere Reserves and Political Skills Transfer in University Curricula

Melki SLIMANI, Angela BARTHES and Jean-Marc LANGE

10.1. Introduction
10.2. Towards a conceptual recontextualization of the political skill in the environmental field
    10.2.1. Learning eco-literacy and building the disposition towards cognitive socialization in academic disciplines

10.2.2. Learning eco-citizenship and building the disposition towards political socialization
10.2.3. Learning environmental deliberation and building the disposition towards critical cognitive socialization
10.2.4. Learning in collective action regimes and building the disposition towards democratic socialization

10.3. Environmental political skill: Master's degree in Man and the Biosphere – case study
10.3.1. The MAB Programme, political skill and formal curriculum
10.3.2. Methodology

10.4. Results and discussion
10.4.1. What political skill in the formal curriculum of the MAB master's degree?
10.4.2. What disciplinary contributions to the environmental political skill in the formal curriculum of the MAB master's degree?

10.5. Conclusion: changing curricular morphologies
10.6. References

Chapter 11. Education and Mediation in the Arganeraie: Alliance Strategies Between Education and Tourism Actors?

Saïd BOUJROUF and Abdullah AÏT L'HOUSSAIN

11.1. Introduction
11.2. Locating the Arganeraie biosphere reserve
11.3. The ABR, a tourist landscape showcased by the media?
11.4. ABR landscape imaging and its dissemination
11.5. A confusion between education forms in the ABR: formal, non-formal and informal
11.6. Towards mediation in the ABR or the construction of an alliance and communication strategies between education and tourism actors
11.7. The territorial integration of the ABR – a condition for the alliance's success: communication, mediation and media coverage
11.8. "Polarized" networks in the ABR: a tool for the alliance between education and tourism actors
11.9. Actor training for the development of capacities: skills and capability for communication management
11.10. Conclusion
11.11. References

Other titles from

in

Science, Society and New Technologies

2023

CARDON Alain, EL HAMI Abdelkhalak
Fundamental Generation Systems: Computer Science and Artificial Consciousness, the Informational Field of Generation of the Universe, the Sixth Sense of Living Beings
(Digital Sciences Set – Volume 4)

ELAMÉ Esoh
The Sustainable City in Africa Facing the Challenge of Liquid Sanitation
(Territory Development Set – Volume 2)

JURCZENKO Emmanuel
Climate Investing: New Strategies and Implementation Challenges

PARET Dominique, CRÉGO Pierre, SOLÈRE Pauline
Smart Patches: Biosensors, Graphene and Intra-Body Communications

VENTRE Daniel, LOISEAU Hugo
Cybercrime during the SARS-CoV-2 Pandemic (2019-2022): Evolutions, Adaptations, Consequences
(Cybersecurity Set – Volume 3)

2022

AIT HADDOU Hassan, TOUBANOS Dimitri, VILLIEN Philippe
Ecological Transition in Education and Research

CARDON Alain
Information Organization of The Universe and Living Things: Generation of Space, Quantum and Molecular Elements, Coactive Generation of Living Organisms and Multiagent Model
(Digital Science Set – Volume 3)

CAULI Marie, FAVIER Laurence, JEANNAS Jean-Yves
Digital Dictionary

DAVERNE-BAILLY Carole, WITTORSKI Richard
Research Methodology in Education and Training: Postures, Practices and Forms
(Education Set – Volume 12)

ELAMÉ Esoh
Sustainable Intercultural Urbanism at the Service of the African City of Tomorrow
(Territory Development Set – Volume 1)

FLEURET Sébastien
A Back and Forth Between Tourism and Health: From Medical Tourism to Global Health
(Tourism and Mobility Systems Set – Volume 5)

KAMPELIS Nikos, KOLOKOTSA Denia
Smart Zero-energy Buildings and Communities for Smart Grids
(Engineering, Energy and Architecture Set – Volume 9)

2021

BARDIOT Clarisse
Performing Arts and Digital Humanities: From Traces to Data
(Traces Set – Volume 5)

BENSRHAIR Abdelaziz, BAPIN Thierry
From AI to Autonomous and Connected Vehicles: Advanced Driver-Assistance Systems (ADAS)
(Digital Science Set – Volume 2)

BÉRANGER Jérôme
Societal Responsibility of Artificial Intelligence: Towards an Ethical and Eco-responsible AI
Technological Prospects and Social Applications Set – Volume 4)

CORDELIER Benoit, GALIBERT Oliver
Digital Health Communications
Technological Prospective and Social Applications Set – Volume 5)

DOUAY Nicolas, MINJA Michael
Urban Planning for Transitions

GALINON-MÉLÉNEC Béatrice
The Trace Odyssey 1: A Journey Beyond Appearances
(Traces Set – Volume 4)

HENRY Antoine
Platform and Collective Intelligence: Digital Ecosystem of Organizations

KOLOKOTSA Denia, KAMPELIS Nikos
Smart Buildings, Smart Communities and Demand Response
(Engineering, Energy and Architecture Set – Volume 8)

LE LAY Stéphane, SAVIGNAC Emmanuelle, LÉNEL Pierre, FRANCES Jean
The Gamification of Society
(Research, Innovative Theories and Methods in SSH Set – Volume 2)

RADI Bouchaïb, EL HAMI Abdelkhalak
Optimizations and Programming: Linear, Non-linear, Dynamic, Stochastic and Applications with Matlab
(Digital Science Set – Volume 1)

2020

ALAKTIF Jamila, CALLENS Stéphane
Migration and Climate Change: From the Emergence of Human Cultures to Contemporary Management in Organizations

BARNOUIN Jacques
The World's Construction Mechanism: Trajectories, Imbalances and the Future of Societies
(Interdisciplinarity between Biological Sciences and Social Sciences Set – Volume 4)

ÇAĞLAR Nur, CURULLI Irene G., SIPAHIOĞLU Işıl Ruhi, MAVROMATIDIS Lazaros
Thresholds in Architectural Education
(Engineering, Energy and Architecture Set – Volume 7)

DUBOIS Michel J.F.
Humans in the Making: In the Beginning was Technique
(Social Interdisciplinarity Set – Volume 4)

ETCHEVERRIA Olivier
The Restaurant, A Geographical Approach: From Invention to Gourmet Tourist Destinations
(Tourism and Mobility Systems Set – Volume 3)

GREFE Gwenaëlle, PEYRAT-GUILLARD Dominique
Shapes of Tourism Employment: HRM in the Worlds of Hotels and Air Transport
(Tourism and Mobility Systems Set – Volume 4)

JEANNERET Yves
The Trace Factory
(Traces Set – Volume 3)

KATSAFADOS Petros, MAVROMATIDIS Elias, SPYROU Christos
Numerical Weather Prediction and Data Assimilation
(Engineering, Energy and Architecture Set – Volume 6)

MARTI Caroline
Cultural Mediations of Brands: Unadvertization and Quest for Authority
(Communication Approaches to Commercial Mediation Set – Volume 1)

MAVROMATIDIS Lazaros E.
Climatic Heterotopias as Spaces of Inclusion: Sew Up the Urban Fabric
(Research in Architectural Education Set – Volume 1)

MOURATIDOU Eleni
Re-presentation Policies of the Fashion Industry: Discourse, Apparatus and Power
(Communication Approaches to Commercial Mediation Set – Volume 2)

SCHMITT Daniel, THÉBAULT Marine, BURCZYKOWSKI Ludovic
Image Beyond the Screen: Projection Mapping

VIOLIER Philippe, with the collaboration of TAUNAY Benjamin
The Tourist Places of the World
(Tourism and Mobility Systems Set – Volume 2)

2019

BRIANÇON Muriel
The Meaning of Otherness in Education: Stakes, Forms, Process, Thoughts and Transfers
(Education Set – Volume 3)

DESCHAMPS Jacqueline
Mediation: A Concept for Information and Communication Sciences
(Concepts to Conceive 21st Century Society Set – Volume 1)

DOUSSET Laurent, PARK Sejin, GUILLE-ESCURET Georges
Kinship, Ecology and History: Renewal of Conjunctures
(Interdisciplinarity between Biological Sciences and Social Sciences Set – Volume 3)

DUPONT Olivier
Power
(Concepts to Conceive 21st Century Society Set – Volume 2)

FERRARATO Coline
Prospective Philosophy of Software: A Simondonian Study

GUAAYBESS Tourya
The Media in Arab Countries: From Development Theories to Cooperation Policies

HAGÈGE Hélène
Education for Responsibility
(Education Set – Volume 4)

LARDELLIER Pascal
The Ritual Institution of Society
(Traces Set – Volume 2)

LARROCHE Valérie
The Dispositif
(Concepts to Conceive 21st Century Society Set – Volume 3)

LATERRASSE Jean
Transport and Town Planning: The City in Search of Sustainable Development

LELEU-MERVIEL Sylvie, SCHMITT Daniel, USEILLE Philippe
From UXD to LivXD: Living eXperience Design

LENOIR Virgil Cristian
Ethically Structured Processes
(Innovation and Responsibility Set – Volume 4)

LOPEZ Fanny, PELLEGRINO Margot, COUTARD Olivier
Local Energy Autonomy: Spaces, Scales, Politics
(Urban Engineering Set – Volume 1)

METZGER Jean-Paul
Discourse: A Concept for Information and Communication Sciences
(Concepts to Conceive 21st Century Society Set – Volume 4)

MICHA Irini, VAIOU Dina
Alternative Takes to the City
(Engineering, Energy and Architecture Set – Volume 5)

PÉLISSIER Chrysta
Learner Support in Online Learning Environments

PIETTE Albert
Theoretical Anthropology or How to Observe a Human Being
(Research, Innovative Theories and Methods in SSH Set – Volume 1)

PIRIOU Jérôme
The Tourist Region: A Co-Construction of Tourism Stakeholders
(Tourism and Mobility Systems Set – Volume 1)

PUMAIN Denise
Geographical Modeling: Cities and Territories
(Modeling Methodologies in Social Sciences Set – Volume 2)

WALDECK Roger
Methods and Interdisciplinarity
(Modeling Methodologies in Social Sciences Set – Volume 1)

2018

BARTHES Angela, CHAMPOLLION Pierre, ALPE Yves
Evolutions of the Complex Relationship Between Education and Territories
(Education Set – Volume 1)

BÉRANGER Jérôme
The Algorithmic Code of Ethics: Ethics at the Bedside of the Digital Revolution
(Technological Prospects and Social Applications Set – Volume 2)

DUGUÉ Bernard
Time, Emergences and Communications
(Engineering, Energy and Architecture Set – Volume 4)

GEORGANTOPOULOU Christina G., GEORGANTOPOULOS George A.
Fluid Mechanics in Channel, Pipe and Aerodynamic Design Geometries 1
(Engineering, Energy and Architecture Set – Volume 2)

GEORGANTOPOULOU Christina G., GEORGANTOPOULOS George A.
Fluid Mechanics in Channel, Pipe and Aerodynamic Design Geometries 2
(Engineering, Energy and Architecture Set – Volume 3)

GUILLE-ESCURET Georges
Social Structures and Natural Systems: Is a Scientific Assemblage Workable?
(Social Interdisciplinarity Set – Volume 2)

LARINI Michel, BARTHES Angela
Quantitative and Statistical Data in Education: From Data Collection to Data Processing
(Education Set – Volume 2)

LELEU-MERVIEL Sylvie
Informational Tracking
(Traces Set – Volume 1)

SALGUES Bruno
Society 5.0: Industry of the Future, Technologies, Methods and Tools
(Technological Prospects and Social Applications Set – Volume 1)

TRESTINI Marc
Modeling of Next Generation Digital Learning Environments: Complex Systems Theory

2017

ANICHINI Giulia, CARRARO Flavia, GESLIN Philippe,
GUILLE-ESCURET Georges
Technicity vs Scientificity – Complementarities and Rivalries
(Interdisciplinarity between Biological Sciences and Social Sciences Set – Volume 2)

DUGUÉ Bernard
Information and the World Stage – From Philosophy to Science, the World of Forms and Communications
(Engineering, Energy and Architecture Set – Volume 1)

GESLIN Philippe
Inside Anthropotechnology – User and Culture Centered Experience
(Social Interdisciplinarity Set – Volume 1)

GORIA Stéphane
Methods and Tools for Creative Competitive Intelligence

KEMBELLEC Gérald, BROUDOUS Evelyne
Reading and Writing Knowledge in Scientific Communities: Digital Humanities and Knowledge Construction

MAESSCHALCK Marc
Reflexive Governance for Research and Innovative Knowledge (Responsible Research and Innovation Set - Volume 6)

PARK Sejin, GUILLE-ESCURET Georges
Sociobiology vs Socioecology: Consequences of an Unraveling Debate (Interdisciplinarity between Biological Sciences and Social Sciences Set – Volume 1)

PELLÉ Sophie
Business, Innovation and Responsibility (Responsible Research and Innovation Set – Volume 7)

2016

ANDRÉ Michel, SAMARAS Zissis
Energy and Environment (Research for Innovative Transports Set – Volume 1)

BLANQUART Corinne, CLAUSEN Uwe, JACOB Bernard
Towards Innovative Freight and Logistics (Research for Innovative Transports Set – Volume 2)

BRONNER Gérald
Belief and Misbelief Asymmetry on the Internet

COHEN Simon, YANNIS George
Traffic Management (Research for Innovative Transports Set – Volume 3)

EL FALLAH SEGHROUCHNI Amal, ISHIKAWA Fuyuki, HÉRAULT Laurent, TOKUDA Hideyuki
Enablers for Smart Cities

GIANNI Robert
Responsibility and Freedom (Responsible Research and Innovation Set – Volume 2)

GRUNWALD Armin
The Hermeneutic Side of Responsible Research and Innovation
(Responsible Research and Innovation Set – Volume 5)

LAGRAÑA Fernando
E-mail and Behavioral Changes: Uses and Misuses of Electronic Communications

LENOIR Virgil Cristian
Ethical Efficiency: Responsibility and Contingency
(Responsible Research and Innovation Set – Volume 1)

MAESSCHALCK Marc
Reflexive Governance for Research and Innovative Knowledge
(Responsible Research and Innovation Set – Volume 6)

PELLÉ Sophie, REBER Bernard
From Ethical Review to Responsible Research and Innovation
(Responsible Research and Innovation Set – Volume 3)

REBER Bernard
Precautionary Principle, Pluralism and Deliberation: Sciences and Ethics
(Responsible Research and Innovation Set – Volume 4)

TORRENTI Jean-Michel, LA TORRE Francesca
Materials and Infrastructures 1
(Research for Innovative Transports Set – Volume 5A

TORRENTI Jean-Michel, LA TORRE Francesca
Materials and Infrastructures 2
(Research for Innovative Transports Set – Volume 5B)

VENTRE Daniel
Information Warfare – 2nd edition

YANNIS George, COHEN Simon
Traffic Safety
(Research for Innovative Transports Set – Volume 4)